Drone Art

The publisher and the University of California Press Foundation gratefully acknowledge the generous support of the Judy and Bill Timken Endowment Fund in Contemporary Arts.

Drone Art

The Everywhere War as Medium

Thomas Stubblefield

UNIVERSITY OF CALIFORNIA PRESS

University of California Press
Oakland, California

Library of Congress Cataloging-in-Publication Data

Names: Stubblefield, Thomas, author.
Title: Drone art : the everywhere war as medium / Thomas Stubblefield.
Description: Oakland, California : The University of California Press, [2020] | Includes bibliographical references and index.
Identifiers: LCCN 2019027395 (print) | LCCN 2019027396 (ebook) | ISBN 9780520339613 (cloth) | ISBN 9780520339620 (paperback) | ISBN 9780520974432 (ebook)
Subjects: LCSH: Drone aircraft—United States—In art. | Drone aircraft—In mass media. | Drone aircraft—Political aspects—United States.
Classification: LCC UG1242.D7 S78 2020 (print) | LCC UG1242.D7 (ebook) | DDC 700/.456—dc23
LC record available at https://lccn.loc.gov/2019027395
LC ebook record available at https://lccn.loc.gov/2019027396

29 28 27 26 25 24 23 22 21 20
10 9 8 7 6 5 4 3 2 1

Contents

Acknowledgments

This book is the product of a six-year conversation that spanned multiple continents and dozens of interlocutors. Working backward through this collection of presentations, publications, film screenings, and exhibitions brings to light the extraordinary generosity of colleagues, friends, and family members, all of whom have helped bring this work into being in unique ways.

The first formal presentation of these ideas took place at the 2013 Dark Side of the Digital conference, hosted by the Center for 21st Century Studies at the University of Wisconsin–Milwaukee. There I was joined by drone scholars Peter Asaro and Patrick Lichty, both of whom shaped my early thinking on the subject with their analytical insight, historical knowledge, and interdisciplinary spirit. In addition to the interactions with my fellow panelists, the exchange between presenters and attendees that took place that day planted a number of seeds that would continue to bear fruit throughout the various iterations of this project. While revising these preliminary ideas for an essay in the *Life in the Age of Drone Warfare* anthology (Duke University Press, 2017), my work benefited greatly from the attention of editors Caren Kaplan and Lisa Parks. As they encouraged me to situate my argument within the broader themes of the publication, I began to expand the scope of the project beyond the confines of art history and media studies. In the process, the prospect of creating a monograph dedicated to this subject began to appear as a possibility for the first time.

It was through subsequent presentations at the University of Southern Denmark's Drone Imaginaries and Society conference and the Medial Reflections: Threat Communication and the US-American Order after 9/11 conference at the University of Tübingen that the central arguments of this book took shape. This is thanks in large measure to scholars such as Svea Bräunert, Derek Gregory, Richard Grusin, Claudette Lauzon, Kathrin Maurer, Jan Mieszkowski, and Lukas R. A. Wilde, all of whom helped to solidify the theoretical foundation of this work and expand the array of objects it engages.

Throughout this process, my colleagues at the University of Massachusetts Dartmouth have been a constant source of guidance and support. Of particular importance to this project were Professors Anna Dempsey, Pamela Karimi, Spencer Ladd, and Rebecca Uchill. In addition, Allison Geremia provided much-needed assistance in manuscript preparation, and Hayden Stubblefield helped with assembling images. I am also immensely grateful for the insight and commitment of my editor, Raina Polivka, and her tireless assistant, Madison Wetzell. Finally, I want to thank Hayden, Zoey, and Damien for supplying a much-appreciated distraction from writing when necessary, and Karen for her unending patience and innumerable sacrifices.

Introduction

The Perverse Symmetry of Drone Art

What happens when a drone enters the gallery or appears on screen? What thresholds are crossed as this weapon of war occupies everyday visual culture? These questions have appeared with increasing regularity since the advent of the war on terror, when the modern drone began migrating into civilian platforms of film, photography, installation, sculpture, performance art, and theater as both subject and medium. Now, nearly two decades later, it is finally possible not only to engage with these issues, but also to articulate the category of "drone art" with some certainty, to outline its primary features, identify its historical lineages, and assess its political aspirations. This task is complicated, however, by the fact that like drone warfare itself—which is defined primarily in terms of its lack of an end point, denial of geographic boundaries, and absence of judicial process—this emerging genre is most immediately legible in terms of what it doesn't do. Not only does drone art avoid explicit condemnation of its subject, but so does it refuse the expectations of protest art more broadly. In fact, next to Picasso's scathing critique of the aerial bombing of civilians in *Guernica*, Anselm Kiefer's cathartic postwar ruins, or the antiwar films of the Vietnam era, these works appear decidedly neutral and aloof.

Not surprisingly, this lack of condemnation or even critique has proved to be a stumbling block for interpreters, many of whom regard the absence of an agenda as a sign of political passivity and/or creative impotence. Consider, for example, the *New York Times* review of Laura

Poitras's drone-heavy *Astro Noise* exhibition at the Whitney in 2016, which begins:

> Political art has changed over the past 50 years. Unlike the protest art of an earlier era, much of the most interesting new work feels slippery and evasive, as if reluctant to speak its mind. In part, this is a reflection of different, though not necessarily evolved, thinking. We've abandoned old beliefs in utopias, in visions—some would say hallucinations—of a society built on absolute good for all, in which art plays a declaratively positive role. The 1960s counterculture, even at its most anarchic, was based on such beliefs and visions. And that counterculture is long gone. . . . In the political thinking of an earlier time, you were either part of the problem or part of the solution, the inference being that there *was* a solution. The art of the present is not so sure.[1]

This passage illustrates the way in which a pervasive nostalgia for moral certainty and the utopian project of Marxism frames contemporary art in terms of loss: the loss of universals, the loss of a coherent system of beliefs, even the loss of the future. If one subscribes to such a position, it is not difficult to see how drone art might be regarded as the personification of this crisis. In the serene skies of Trevor Paglen's high-magnification photographs of drones in flight, the empty silhouettes of Predator drones that adorn city streets in James Bridle's *Drone Shadows* project, the anticlimactic appearance of pushpin icons to designate the location of drone strikes in Josh Begley's Metadata+ app, the decontextualized sounds of drone engines in *A Study into 21st Century Drone Acoustics*, and even the banal monotony of suburban life in drone films such as *Good Kill*, one senses a recurring affective detachment and overall lack of antagonism toward the violence of drones. On this basis, drone art would appear to exemplify what the editors of the journal *October* describe as a generalized "absence of visible opposition within the milieu of cultural producers working in the sphere of contemporary visual culture."[2] However, in situating this work in relation to the technological and operational sphere of martial networks, a more nuanced reading emerges, one in which the apparent passivity of this genre is not only a conscious response to the specific conditions of drone power, but in fact the means for reimagining its relations of violence.

While the nonprogram of drone art is necessitated by a number of historical, political, and technological factors, perhaps the most immediate is the increasingly incestuous relationship between martial networks and civilian media. This narrative is most often associated with the realm of video games where the "PlayStation war" takes on an almost literal dimension: Raytheon's universal remote for robotic vehicles uses the graphical user interface (GUI) of *Halo* and *Splinter Cell*; the Wiimote provides the "joystick" for the army's improvised explosive device (IED)-disarming drones; the PlayStation controller is used for the land drone "Dragon Runner"; and the early control system for the Predator drone ran on an Xbox processor. However, the case of video games is hardly unique in its disregard for the boundaries between civilian media and drone technology. In fact, inside the Conex boxes at Creech Air Force Base, the analysts, operators, and legal teams of the kill chain communicate with a program that mimics the functions of Yahoo Messenger. The tagging features of the data collection programs of drone operation were derived from the software used in live National Football League broadcasts. The Rand Corporation even integrated the logging protocols of reality TV shows such as *Jersey Shore* into its algorithms for sorting the enormous amounts of footage generated by drones.[3] Most recently, Google's algorithms were used to catalog drone footage as part of a controversial program called Project Maven.[4] As the rise of the commercial drone industry confirms, the integration of everyday media into military applications is also inseparable from an inverse trajectory, in which the latter are just as often folded back into civilian platforms. From this overlapping history a shared ground of operations emerges that imbues everyday media such as the mobile device with an inconceivable simultaneity, such that the capacity to serve as both medium of civilian communication and drone targeting system is not only possible, but in fact comprises the very conditions of everyday mediality.[5]

As this entanglement between military and civilian media increasingly eradicates an outside from which media-based critiques might be staged, "protest art" and its attendant gestures of resistance and negation prove problematic. And yet it is precisely such responses that the drone's display of state power seems to elicit. This contradiction is a product not only of the nostalgia for the political movements of the previous century that Holland Cotter notes in his *New York Times* piece, but also of the unique structure of drone power itself. Drones articulate an authority that is distributed and elusive, wrapped up with the horizontality of networks, which they materialize as force. At the same time, this diffuse organization of power coexists alongside and at times congeals into more familiar centralized models. In this, drones exemplify what Alexander Galloway and Eugene Thacker describe as the critical juncture between sovereignty and networks, that irresolvable dialectic that is at the core of understanding the operations of power in the twenty-first century.[6] As this modulated duality allows drone power to work through a dynamic array of objects and actors, the enduring presence of vertical hierarchies within these formations pose something of a red herring, perhaps even a lure for the critical project, as they prompt familiar strategies of negation that drone warfare fundamentally eludes.

The ability of the distributed networks of drone operation to sabotage critical interventions is reinforced by both the public discourse of "surgical precision" and the asymmetrical nature of drone operation. The absence of American casualties resulting from drone attacks means that the historical fail-safe of public resistance fails to hold this power in check from without, and as a result the military is more willing to engage targets than it would be in the context of traditional modes of combat. In this way, drone power spreads across national boundaries and obliterates durational thresholds of war, all the while producing large numbers of civilian deaths that are strategically elided through the discourse of targeted killings. With narratives of securitization presenting global violence as a zero-sum game, an either/or proposition inscribes these relations so that one is forced to choose between casualties taken on America soil via large-scale terrorist attacks on civilians or casualties confined to the other side of the war, where targets appear to be chosen more judiciously. In this logic, the nationalist undertones of imperialism become increasingly indecipherable from humanitarian discourse, leading to the possibility of an "ethical defense" of drone strikes.

The self-sabotaging quality that resistance can take on in this environment is perhaps best illustrated through the practice of drone hunting.[7] Associated with the American West, the grounding of drone flights by armed private citizens is often portrayed as a disruptive return of the gaze that overturns hierarchies in the course of restoring individual autonomy. In this view, the practice grants the subject an agency that is charged with negating those governmental intrusions that conflict with self-determination, despite the fact that the same discourse all too often understands drones as part of a globalized counterinsurgency that serves as the necessary prerequisite for such freedoms.[8] As such, these displays of "autonomy" comprise a relation of "cruel optimism," which reaffirms not only the dispersed networked status of the drone but also one's interpolation within its networks.[9] The temptation of critical or artistic projects to slip into a demythologizing mode that seeks to unmask the authority behind the eye of the drone camera, to demystify and destroy the false consciousness produced by its apparent invisibility, carries with it the possibility of a similar kind of contradiction. In positing drone power in terms of the hierarchies of sovereignty, such critiques enact a kind of rhetorical drone hunting that may not only miss the target, so to speak, but actively obscure its dual character. As these operations call upon an imaginary outside from which to issue their critiques, they in fact threaten to reproduce the conditions of what Derek Gregory has called "the everywhere war," confirming that the "old beliefs in utopias" and their concomitant categories of periphery and center are not simply ineffective, but actively appropriated by drone power.[10]

Reading drone art in conjunction with the specific conditions of drone power and the impasse they create for political art challenges the knee-jerk tendency to regard these works as disengaged or even indifferent to the violence of their subject. By rejecting the enduring categories of resistance and critique, the basic premises of "protest art" for more than a century, drone art does not so much evade the politics of its subject as actively reproduce and at times even operate within the mode of power that unmanned aerial vehicles (UAVs) articulate. Staking out this perversely affirmative relation to network-centric warfare ultimately provides the conditions to redirect, reimagine, or otherwise introduce interference from within these relations. However, unlike much of the tactical media interventions of the twentieth century, this second-order operation, by which the mirroring operation of drone art slides into a kind of manifest criticality, is always already external to the work itself. It exists only as that which is to be added by users, accrued through circulation, or otherwise brought into being by inorganic agencies within a larger media ecology. As such, the transgressive gesture is necessarily left unarticulated by the work itself, which offers only the vague insistence that an alternate realization of drone warfare is latent to the specific forms, network relations, hierarchies, and histories of its operation.

From this perspective, the apparent passivity of drone art is the surface feature of a dialectical intervention by which this detachment shifts strategically into its other. This possibility is a product of drone art's affinity with the "post-Internet art object," which elicits collective or distributed modes of authorship via digital networks in order to undercut the singularity of the work with successive iterations. From this perspective, the danger of this nonprogram is not so much its failure to engage with power, but rather its capacity to engage too intimately. After all, in pursuing these relations, the genre necessarily treads a fine line between the utopian potential of the network and the reaffirmation of a dispersed mode of power, which Michael Dillon and Julian Reid refer to as "martial becoming."[11] Indeed, the latter concept internalizes the slipperiness of this distinction by testifying to the fact that "the very means by which phenomenality is itself engendered—the intimate correlation between appearing and what appears, enacted through the power of signification—has become, above all, the prowess to which martial embodiment, paradigmatically represented by the United States military, now aspires."[12] As drone art attempts to oscillate between these designations, it creates the very real prospect that its interventions will simply reiterate, even bolster the operations of drone warfare. Such is the gamble of this genre, which seeks to maintain what Bruno Latour refers to as the "reversibility of foldings" by way of a kind of self-conscious acquiescence to the very relations it seeks to upend.[13] The critical questions then become: What differentiates these relations of perverse symmetry from simple reproduction? What allows such formations to reimagine power outside of existing formations? It is in this regard that accelerationism forms an important interlocutor with the genre of drone art.

ACCELERATIONISM AND THE NONPROGRAM OF DRONE ART

While referencing a diverse set of theories, which include Nick Land's absolute deterritorialization, the epistemic approach of Ray Brassier and Reza Negarestani, and Gilles Deleuze and Félix Guattari's schizoanalysis—accelerationism begins with the recognition that there is no longer any outside to capitalism (a position that maintains a particular resonance with the global counterinsurgency), and therefore the primary way to derail its inner workings is by amplifying its own powers for deformation and self-destruction.[14] As a political philosophy, accelerationism spawns considerable disagreement regarding what exactly should be accelerated and whether this process requires outside intervention or is simply the natural end point of capitalism itself. Without downplaying the importance of this debate, it is safe to say that these theories seem to have found more coherent expression in the realm of aesthetics, so much so that theorists such as Steven Shaviro have insisted that accelerationism be regarded primarily as an artistic program rather than a political economy.[15] The compatibility between accelerationism and contemporary art practices is evident in the following remarks by

Robin Mackay and Armen Avanessian: "While distancing itself from mere technological optimism, contemporary accelerationism retains an antipathy, a disgust even, for retreatist solutions, and an ambitious interest in reshaping and repurposing (rather than refusing) the technologies that are the historical product of capitalism."[16] While this emphasis on "reshaping and repurposing (rather than refusing)" clearly resonates with the art movements of the twentieth century, it maintains particular relevance to the situation of drone art. As the accelerationist impulse "is already implicitly present in the speculative re-application of drone technology," drone art fully embodies the counterintuitive nature of an accelerationist program, which reproduces that which it abhors in order to escalate the latent formations that operate within this larger ecology of actors.[17] In rejecting the possibility of a symbolic intervention, this engagement is defined in primarily structural terms. As a result, its reproduction of drone power is easily mistaken for the cool detachment of postmodern pastiche. However, as drone art is driven by a nonprogram rather than an antiprogram, its rejection of traditional modes of critique and resistance is not synonymous with the rejection of the political as a category or of the possibility of intervention. In fact, it is precisely by way of these relations of symmetry that this work seeks to amplify the internal contradictions and asymmetries of drone power, to harness the relations of becoming that inhabit this network in order to realize alternate iterations of its force. Though drone art deploys a wide variety of practices and strategies, its operation primarily proceeds according to three recurring modes: the networked, the objectile, and the historiographic. In each case, the efficacy of this engagement is contingent upon not simply reproducing the structures of drone power, but internalizing the dialectic relationships that comprise its functioning.

The genre's relationship with the civilian networks of social media, news aggregators, navigation software, GPS satellites, and mobile apps is central to understanding its political project. As the remote nature of drone power's asymmetry ensures that its violence is spatially displaced for operators and constituents, the enfolded martial presences that populate these everyday platforms represent critical points of access to an "elsewhere war." To draw out these latent formations, drone art understands digital networks as not simply vehicles for disseminating content, but active agents capable of charting the otherwise inaccessible collaboration between martial and civilian networks. Exemplary in this respect are James Bridle's *Dronestagram*, which uses Google Maps and the popular image app Instagram to catalog drone strikes, and the

#NotABugSplat project, a monumental portrait of a victim of drone strikes that entangles the eye of the UAV with the sharing economy of social media. As these works enlist the human users and inorganic agencies of these platforms to manifest the tensions of drone power, they attempt to provide the conditions by which it becomes possible to, as Ulises Mejias puts it, "unthink the network."[18]

In those instances when the network is not the primary means of acceleration, the artifacts of drone art internalize these relations by making use of the pliable and intermedial nature of digital media, a condition that Deleuze associates with the "objectile."[19] As the kill chain displaces the stasis of representation with aggregated conditions and predictive models, drone power articulates a field of weaponized becoming in which proper nouns serve as a symptom as much as an origin of violence. Interpenetrating platforms such as 3D printing, photography, installation, and video, the assets of drone art, move across iterations, forming a chronology of shape-shifting that echoes this abandonment of identity as the anchoring determinant of martial operations. The ontological instability that ensues from these relations is evident in works such as Nathaniel Hartman's *ASM_frag* (2014), a sculptural fragment born out of an extensive chain of iterations that range from an online photograph of a missile shard from a US drone strike to a report of this artifact having been smuggled out of Afghanistan. Matthew Fuller, Graham Harwood, and Matsuko Yokokoji's *Endless War* (2012) installation performs the constitutive breakdown of representation of drone power by projecting leaked dispatches of the kill chain according to "repeated patterns of events, spatial information, kinds of actors, timings and other factors."[20] In sorting these data via alternate parameters of an N-gram analysis, the work draws upon both the varied modes of address of installation and the becoming of the postrelational database in order to not just reanimate these artifacts, but conceive of new totalities that they might operate within.

In addition to employing these media-centered strategies, drone art introduces alternate histories of the UAV to exacerbate the ontological tensions that reside within its interfaces and procedures. Central to this enterprise is the animal prehistory of drones, which is regularly called upon not only to unravel the progressivist technological narratives that inform drone discourse, but also to draw out the residual animal presence that haunts their operation. For example, Mato Atom's short film *Seagulls* uses intertextual relations with familiar fairy tales and formal techniques such as compositing and non-diegetic sound to push this historical connection between the drone and the nonhuman animal to the point of indecipherability. While Mato's work stages this intervention within the technical substrate of the moving image, Dutch artist Bart Jensen's zoo-drones perform the animality of the drone in more literal terms. Using UAV technology to give life to taxidermied animals, the uncanny flight of these hybrid creatures refuses simple colonization of the animal body and instead produces a movement that undercuts identity and destabilizes the agencies that populate this media ecology. As the nonhuman animal and the drone collapse into one another in these historiographic works, not only does the linear quality of drone history come undone, but so do the coordinates of its violence. In this way, the animality of this genre posits drone history as a dynamic and reversible field in order to open the future of this technology to equally unforeseeable iterations.

These networked, objectile, and historiographic modes of drone art proceed according to processes of intensification. Eschewing an explicit program, position, and at times even discernible content, they offer the means of recuperating unformed possibility from within systems of control rather than delivering specific or ideal iterations of the media ecology, discourses, and politics that inform drone operation. In echoing the horizontal and distributed nature of drone power, this genetic aspect of drone art discloses the capacity of martial networks to break down the parameters of asymmetry that determine their violence. This is routinely dramatized in drone film and theater, where geographic ambiguity, temporal slippages, and unsettling plot symmetries that elide the distinction between "us" and "them" do not simply "humanize the enemy," but eradicate the position of safety from which citizens of the Global North often experience drone power. As the Arizona desert, the shopping mall, and the microdramas of suburban life lose their distinctness from targets overseas, the groundlessness of drone power offers both the prospect of boundless violence and the becoming-other of unformed potentiality. With all of this in mind, it seems that Karen Beckmann's description of the work of drone artist Trevor Paglen as an "activism . . . founded on ambiguity, incomplete understanding, doubt and obscurity, rather than slogans, unity, loyalty and coherence" may very well extend to the genre as a whole.[21] And yet this statement must come with some qualifications. As this "activism" posits deformation as the limits of its intervention, it necessarily stops short of the kind of advocacy that this term typically embodies. In the place of the oppositional message, drone art installs a strategic deferral, which calls upon the organic and inorganic agencies of networks, the iterative capacity of digital media, and the nonlinear histories that animate its technologies to reimagine drone operation outside of its existing program.

Tracing this narrative requires that both terms of the phrase "drone art" be understood as expansive and co-constitutive categories. As the material correlate to geopolitical relations, nationalist histories, and racial discourses, drones clearly exceed the parameters of a purely technological object. While such convergences of material and immaterial forces comprise the boundaries of all machines, drones present an especially comprehensive and dynamic instance due to the distributed nature of the kill chain and the remote relations it reproduces. As these attributes preclude the possibility of isolating "the drone" at technological, phenomenological, or symbolic levels, the familiar white silhouette can never be wholly collapsed with "drone." The latter is only a fleeting representative of a fluid system of relations that manifests through objects. Its presence is nondefinitive in itself, a residue of the perpetual elsewhere that results from what Arthur K. Cebrowski and John J Garstka describe as "network-centric warfare."[22] By emphasizing information "pull" rather than the "push" of broadcast, this decentered model of military operations flattens cumbersome hierarchies of chain of command so as to enhance the exchange of information, increase overall flexibility, and accelerate responsiveness in the face of rapidly changing conditions of warfare. In the process, data flow among networked censors, human agents on the ground, analysts, and commanders horizontally and in near real-time via satellites and high-speed fiber optic cables. To acknowledge the dispersed quality of this object of study, this project understands drones as not only an ecology of weaponized media, but the myriad of processes that materialize the virtual relations of martial networks as force. From this expanded field, art history, media studies, surveillance studies, geography, animal studies, critical theory, and military history become critical tools in excavating the deep convergences of the drone's dynamic martial folds.

METHODOLOGY WITHOUT MELANCHOLY

Taken in isolation, the word *art* would appear ill-suited for a body of work that includes not only photography, installation, and painting, but also experimental music, phone apps, theater, and film. However, when paired with a military technology this word introduces a productive conflict that succinctly conveys the larger stakes of this work. As the response to early presentations of this project has made clear, the persistent notions of beauty, critique, and resistance that accompany this designation present the possibility of "drone art" as something of a contradiction in terms. It is as if the phrase lacks a much needed conjunction to preserve the integrity of these fields (perhaps "drones and art" or "drones in art?"). However, it is for this reason that the shortened designation *drone art* proves productive, as it performs the perverse symmetry that characterizes this genre by allowing the role of modifier to move seamlessly between referents. The result is an ambiguous directionality that undercuts the solidity of both terms: Does *art* refer to artifacts created by drones? Does it suggest that drone operation maintains artful qualities? Or does it imply that a given set of works displays drone-like attributes? Read in conjunction with the genre's mirroring function, the answer to these questions is of course "yes," as collapsing such oppositions is the lingua franca of both drone power and the genre of work that engages it.

From this perspective, *drone art* is not simply a descriptive term, but a methodology. It speaks to a mode of inquiry that begins within a conflicting field of symmetry in which military doctrines such as network-centric warfare are read as aesthetic programs, and everyday social media platforms are analyzed as weapons of war. By embracing the internal contradictions (and blasphemies) of this genre, such an approach disentangles drone art from what Wendy Brown calls "left melancholy," a turn-of-the-century phenomenon in which an enduring allegiance to "notions of unified movements, social totalities and class-based politics" forecloses the possibility of an alternate, equally transformative engagement with the present.[23] This discourse, in which the attachment to the object of loss seems increasingly to "supersede any desire to recover from this loss," posits "the sixties" as a historical absolute against which alternate modalities of political art can only appear as lapses or failures. The methodology of drone art jettisons this position in order to engage the critical impasse posed by this new mode of power, an impasse that Caren Kaplan succinctly conveys when she asks: "How do you produce critique when various entities that appear to be distinct are, in fact, quite closely related?"[24] Such a prompt might best be understood as a rhetorical question, as the nonprogram of drone art resists the possibility of critique. Instead, the writing of drone art necessarily takes place within the enfolded relations of drone symmetry, forming a methodological chiasmus in which thinking drone power through art and thinking art through drone power prove to be inseparable tasks.

This book is organized around a series of recurring conceptual knots or convergence points that inform the nonprogram of drone art. In order to build the foundation necessary for engaging these entangled histories, theories, and practices, chapter 1 presents an overview of drone power. Of particular importance to this discussion is the "signature strike" which forges targets according to a "disposition matrix" of behaviors rather than the specific identity of the subject. With this emphasis on plausible identities, meanings, and events, the role of martial networks shifts from producing a definitive world picture to overseeing the ground from which such representations emerge. The flattened ontology that emerges from these relations allows drone power to penetrate the world directly, to work *through* and *as* instead of *upon* its objects. These operations form the basis for the complementary logics of drone power, which I call "tactical animism" and "simple triggers." Framing drone power in these terms reveals a shared logic at work in Trevor Paglen's appropriated drone video piece *Drone Vision* (2010), Joy Garnett's *Predator* paintings (2011–2012), Noor Behram's photographic exhibition *Gaming in Waziristan* (2007–2011), and Laura Poitras *Bed Down Location* (2016) installation, which in distinct ways attempt to internalize this logic of the signature strike so as to reclaim and/or redirect the drone's capacity for world-making.

Despite its undeniably photogenic quality, the drone maintains a complicated relation to the still image as a result of this capacity for world-making. Not only does the singularity of drone presence betray the dispersed nature of this mode of power, but so too do its relations of martial becoming appear ill-suited to photography's emphasis on stasis, iconicity, and authoritative representation. Chapter 2 pursues this conflict from within the kill chain, where representational images are interpenetrated by an immense archive of diagrammatic signs that includes heat signatures, chemical compositions, and GPS coordinates. This "mixed semiotics" does not so much deliver static conditions as build instrumentalized diegeses around an impending attack. In the process, the representational matrix of ocular media gives way to what Mike Hill calls "fractal media," mathematical simulations that "rethink" the profilmic event rather than visually represent it.[25] Works such as James Bridle's *Drone Shadows*, Trevor Paglen's *Untitled* series, Mahwish Chisty's *Kill Box* (2016), and Scott Patrick Wiener's *Visibility Tests for Sky* (2011–2015) attempt to reproduce these relations by rendering the photographic frame similarly porous. In opening the still image to external flows that undercut its ontological stasis, they reiterate an operational schema of drone warfare called "the kill box." Critical to this prospect is the ability of drone art to integrate the distributed networks of social media, the iterative nature of digital images, and the dispersion of authorship these platforms engender.

While relations of invisibility and "seeing without being seen" comprise critical components of drone operation, the recurring reliance upon panopticism in theorizing its mode of power proves problematic. Not only does this conceptual framing tend to obscure the nonvisual elements that contribute to the drone's mode of address, but so too does it overlook the gestures of revelation that prove equally critical to its operation. By eliding the often traumatic modes of presence that drones articulate for subjects on the ground, such readings can easily universalize the position of safety of those untouched by this violence and thereby reproduce the asymmetry of drone operation at a discursive level. In order to pursue a more nuanced understanding of drone presence, chapter 3 begins with firsthand accounts of those living under drones.[26] From this perspective it becomes clear that conventional dynamics of surveillance do not simply coexist alongside an opposing logic of explicit presence, but drone power in fact relies on a strategic oscillation between panoptic modes of invisibility and overt revelation. Drone art proves useful in codifying these relations of "post-panopticism" as it often foregrounds the embodied experiences of those below over the omniscient aerial view of the UAV itself. Exemplary of this dynamic are James Bridle's *UAV Identification Kit* and Ruben Pater's *A Study in 21st Century Drone Acoustics*, which mobilize the civilian practices of "spotting" used during World War II and the military history of ambient sound recording to undercut the stability of referents that maintain drone warfare's asymmetry. In similar fashion, Bridle's *Dronestagram* and Josh Begley's Metadata+ app utilize the locative capacity of the mobile device to interject drone violence into the everyday and thereby unsettle the geographic distinctions that structure its martial networks. Collectively, these works present a reciprocal gaze that forcibly relocates the drone's aerial modes of invisibility to the ground, where its violence extends beyond the existential threat of the strike and into the everyday sensory experiences of its subjects.

As a result of the historical reliance of war on the animal body, modern technologies such as the drone are haunted by a lingering zoological agency. Of the nonhuman animals that occupy the networked bestiary of the drone, none are as influential as birds. Not only do these creatures offer a means of materializing an enduring connection between flight and surveillance that had occupied the imaginary for centuries, but they also introduce the time-space differential of an aerial proxy by which a thoroughly modern asymmetry of warfare would come into being. Chapter 4 attempts to expand the inventory of agencies that comprise the kill chain by excavating this avian prehistory of the drone. Critical to this narrative is the displacement of motor and ocular functions to the animal via Dr. Julius Neubronner's experiments with pigeon photographers during World War I and the outsourcing of processes of targeting to pigeons in B. F. Skinner's World War II Project Orcon. In the films, installations, and sound recordings of drone art, this lingering presence of animality provides a point of entry for opposing modes of convergence. The first, represented by Mato Atom's short film *Seagulls* (2013), seeks to amplify the ontological slippage of these relations between animal, human, and machine so as to reimagine the hierarchies that subtend these terms. The *#NotABugSplat* installation in Pakistan exemplifies an opposing trajectory, which centers on reclaiming the category of the human from these fields of interspeciality by way of a strategic regression toward stasis, specificity, and identity. Taken collectively, these tactics comprise complementary operations that reproduce the animal assemblages of drone power in order to reimagine the parameters by which targets come into being.

As drones appear in film and theater, their dual status as object and medium creates the conditions for a unique mode of reflexivity. Here the remote sensing of the kill chain manifests as a distributed mode of appearance in which the drone is never simply the subject of the image or autonomous material presence, but also partial origin of the diegetic world in which it appears. The deterritorialization that ensues from these relations encompasses both the diegetic and/or institutional boundaries in which the work is presented and the medium from which the presentation takes place. In this way, these platforms come to articulate a contingent and contradictory mode of presentation that is not unlike the "differential specificity" that Rosalind Krauss ascribes to twentieth-century art practices.[27] Chapter 5 traces these relations through the film *Good Kill* (2014) and the theatrical production *Grounded* (2013). While the ability of the drone to achieve full presence on the screen or stage may be contingent upon an active interpenetration of the host media by the larger processes of network-centric warfare, these works demonstrate the myriad of conflicts such a proposition introduces. In the case of *Good Kill*, the automated, collective, and remote nature of drone operation clashes with both the underlying visual relations of narrative cinema and the conventions of the war film genre. In *Grounded* the infiltration of the stage by the drone sacrifices the Aristotelian unities to what Francois Lyotard describes as a "general desemiotics" of theater.[28] However, both works integrate the creative capacity of the kill chain into their respective modes of presentation by way of these conflicts. With this gesture, they invite drone power to work upon the medium itself so as to reimagine the mode of critique by way of its object.

Pursuing these topics reveals a genre forged on the one hand by a skepticism toward representation, resistance, and critique and a strategic embrace of the modulated becoming, distributed collectivities and dynamic folds of the drone on the other. A decade ago William Gibson's novel *Spook Country* (2007) described the conditions that would give birth to this conflicted position. In this work Bobby, a reclusive programmer who makes his living by helping young artists realize their vision with geospatial technology while troubleshooting navigation systems for the US military on the side, explains: "The most interesting ways of looking at the GPS grid, what it is, what we do with it, what we might be able to do with it, all seem [. . .] to be being put forward by artists. Artists or the military. That's something that tends to happen with new technologies generally: the most interesting applications turn up on the battlefield, or in a gallery."[29] While this unsettling structural affinity remains relevant to contemporary drone power, in the years since Gibson's novel the "or" of this proposition and its insinuation of separate and distinct spheres has given way to an inexhaustible and all-inclusive "and." Within these enfolded relations, the "old beliefs in utopias" and their attendant oppositional strategies lose their currency and become self-sabotaging modes of affirmation. And yet these efforts are ineffective only to the extent that alternate realizations of this mode of power are immanent to its operations. As a result, the exploit of drone power is always internal to its own networks, practices, and discourses. It is this realization that structures the varied nonprograms of drone art.

CHAPTER 1

Signature Strikes and World-Making

On March 17, 2011, a US drone fired four Hellfire missiles into a public bus depot in Datta Khel, a town in the Waziristan border region of Pakistan. The attack was not prompted by the presence of a suspected terrorist or other member of the "kill list." In fact, the identities of the subjects on the ground were unknown to the drone pilots, image analysts, and sensor operators of the kill chain that day. Rather, the incident was an example of a "signature strike," in which targets are identified according to a "disposition matrix" of behaviors, locations, and personal relationships. The public is perhaps most familiar with this logic via the practice of the "double tap," in which those who come to the aid of a victim of a drone attack or attend the funeral of a suspected terrorist are themselves targeted by virtue of their presumed connection to the deceased or injured. However, intelligence analysts confirm that signature strikes are more often prompted by relatively banal scenarios such as a typically busy bridge that suddenly empties, a border crossing conducted at night, or possession of a cell phone used to contact a person of interest.[1] In the case of Datta Khel, the triggering event was a meeting of tribal of elders that was held to solve a local dispute over a chromite mine. Despite the fact that the elders had informed the Pakistani army of the meeting in advance, the event was read as a tactical formation by the US military. As a result, forty-four people died, all of whom are believed to have been innocent noncombatants. While the government continues to frame such events as "targeted killings," sources report that the majority of drone attacks conducted by the US military are now based on these "patterns of life" calculations rather than the identification of specific targets.[2]

Officially, such strategies are explained as the necessary correlate to terrorist networks themselves, which tend to be nonhierarchical, loosely defined, and dynamic in nature.[3] In response to these ad hoc formations, it would only seem logical that the operations of the Department of Homeland Security prioritize "the value of the network structure rather than the characteristics of the individual."[4] However, such assessments not only underestimate the consequences of this intervention but also mislead the public as to the true nature of these operations. The exercise of drone power is not simply a matter of utilizing individual actors or events to get at the root causes of terrorism, to work backward from the singular agent toward the larger social field that produces this violence. Nor can its interventions be understood as an acceleration of the familiar operation in which intelligence gathered from a generalized field is used to locate and neutralize a specific actor. Rather, drones seeks to actively disconnect identity, places, and events from their carriers so as to integrate the ground from which proper nouns emerge into their operations. This possibility is contingent upon the real-time circulation of diagrammatic signs (data from sensors, algorithms, and other inorganic assemblages) and more familiar modes of representation (images, verbal and textual descriptions), which together provide the conditions under which targets come into being as such.

As the architectural *dispositifs* of the disciplinary society began taking on a new fluidity at the end of the last century, Mark Poster announced the arrival of a "superpanopticon" in which decentralized iterations of power effectively merged with the everyday via digital networks. Critical to this expansion was "the multiplication of the individual, the constitution of an additional self, one that may be acted upon to the detriment of the 'real' self without that 'real' self ever being aware of what is happening."[5] As the currency of this "data double" was derived from its direct correlation with the "true" self, primary computational signifiers such as Internet protocols (IPs), email addresses, SIM cards, and phone numbers came to serve as carriers around which the metadata harvested from behaviors, locations, and communications could be made not only legible, but actionable. As a result, the superpanopticon articulated a contradictory modality of power in which identity, however partial or diagrammatic, remained partially external to the very processes of modulated becoming that allowed the surveillance apparatus to be absorbed into daily life.

As the martial networks of drone power escalate the distributed nature of these operations, this central tenet of the superpanopticon no longer holds. Rather than "multiplying" the subject via the creation of data doubles, the kill chain correlates identity with a fleeting collection of signifiers that is no longer bound by the transcendent ground of its panoptic predecessor. In this way, it creates the conditions under which subjects/targets are made rather than found. This scenario is confirmed by a recent study conducted by the UK-based nonprofit group Reprieve, which found that each individual on the "kill list" were killed an average of three times by drones.[6] As a result of these relations, drone power not only empties the subject of its singularity but also annihilates the temporal limitations by which it is typically confined. The strategic mode of becoming that underlies this operation intertwines a futural vector in which probable outcomes validate the preemption of a strike with a backward-looking, posthumous trajectory. Identity is thus conferred upon "bare life" in order to retroactively necessitate its extermination.[7] The entanglement of these trajectories allows martial networks to transcend the horizon of the living present. Not only is death no longer the limit case of its operations, but its arrival in fact provides the condition of possibility for its violence. With the target of drone power enveloped by these modulated processes of becoming, the question then becomes: What operative value circumscribes the martial networks of drones? After all, the recuperative function of this network, the process by which sovereign power is ultimately reaffirmed by this radical immanence, is unimaginable without the persistence of some a priori idea, event, or structure. Otherwise the kill chain would produce a pure and unfettered becoming whose radical unforeseeability would serve no master. The answer is that martial networks replace the identity of the subject as the anchoring determinant within this field of difference with the strike itself. As a result, drone violence forms a latent presence that preexists the subjects necessary for actualization. It is, so to speak, given in advance.

In pursuing *a* target rather than *the* target, drones posit the world in terms of an undifferentiated field of inorganic and organic agencies. As the objects, topographies, places, and events that appear within the kill chain are subjected to the same processes of becoming that produce the human subject, a "flattened ontology" emerges that allows drone power to penetrate the world directly, to work *through* and *as* instead of upon its objects. This process echoes the democratizing spirit with which object-oriented ontology has approached the inanimate sphere, in that its "flat" organization similarly insists upon what Levi Bryant describes as a "reject[ion] of any ontology of transcendence or presence that privileges one sort of entity as the origin of all others and as fully present to itself."[8] Indeed, the primary objective of this operation of drone power is not to produce authoritative representations or even hypothetical simulations, but to automate the symbolic realm to such an extent that subjects, landscapes, and events appear to self-generate identities that necessitate a strike. As such, drones do not necessarily take over the human functions of targeting, but rather impregnate the material world with a horizontal field of potentialities and triggers from which attacks materialize as inevitabilities. On account of these relations of "autogenesis," both human subjects and the inanimate sphere they occupy are shot through with a life-giving force that paradoxically necessitates their eradication.[9] These operations comprise the complementary logics of drone operation, which I call "tactical animism" and "simple triggers." Through this joint constellation, drone power shifts from the symbolic to the ontological; its intervention becomes one of world-making rather than meaning-making.[10] Most immediately, world-making acknowledges the drone's control over an expansive milieu, which extends beyond individual subjects to include land features, material bodies, and the host of prepersonal events that are translated into familiar narratives of the global counterinsurgency. However, the term also bears witness to both the exhaustive scale and productive nature of the kill chain's relationship to those external conditions, relations of causality, and symbolic systems that necessitate an attack. Collectively, these entanglements push drone power beyond the historical practices of targeting toward a more constructivist relationship to the world it encounters.

As drones migrate from spaces of conflict to the gallery, screen, or stage, these relations of world-making resurface within everyday media. Even if the UAV itself is not physically present in a given work, its production of the artifacts on display creates an implied presence that recodes these images, installations, or videos according to this constitutive movement between object and origin of the gaze, artifact, and agency. Through this movement, drone art betrays a distinct historiography, a mode of telling that refuses the solidity of its own narrative in order to both capture and reimagine its subject. To begin charting this narrative, the first half of this chapter outlines the concepts of tactical animism and simple triggers in conjunction with readings of Trevor Paglen's appropriated drone video piece *Drone Vision* (2010), Joy Garnett's *Predator* paintings (2011–2012), Noor Behram's photographic exhibition *Gaming in Waziristan* (2007–2011), and Nathaniel Hartman's sculptural artifact *ASM_frag* (2015). This discussion then provides the basis for a consideration of the role of data, databases, and algorithms in drone power, a topic taken up by two works of installation art: Matthew Fuller, Graham Harwood, and Matsuko Yokokoji's *Endless War* (2012) and Laura Poitras's *Bed Down Location* (2016), both of which conjure the drone's processes of world-making by prompting the viewer to effectively perform its constitutive breakdown of representation.

SIMPLE TRIGGERS AND TACTICAL ANIMISM

While the dystopian prospect of autonomous killer robots figures prominently in the collective anxiety regarding drones, the dangers of this technology might just as easily be described according to the exact opposite scenario, i.e., the all too human nature of drones. In this regard, distinguishing the automated processes of targeting that occur via the martial networks of drone warfare from what is typically considered to be the more thoughtful human modes of sighting is perhaps not as easy as we might like it to be. Not only is the latter always already prosthetic and therefore unisolatable, but so too is its predictive aspect structurally related to the processes by which the kill chain translates perceptions into force. Consider William James's discussion of identifying and striking targets in his 1890 *Principles of Psychology*: "A sportsman, while shooting woodcock in cover, sees a bird with the size and color of a woodcock . . . but through the foliage, not having time to see more than that it is a bird of such a size and color, he immediately supplies by inference the other qualities of a woodcock, and is afterwards disgusted to find that he has shot a thrush."[11] Curiously, the passage omits the firing of the gun as well as the deliberation of the subject from this scenario, shifting instead directly from the misrecognition of the hunter to the corpse of the dead bird. This ellipsis imbues the causality of "afterwards" with a productive ambiguity. While one might reasonably assume that the preposition applies to the firing of the gun, semantically speaking, it references the misrecognition of the hunter. As a result, the visual inference of the hunter becomes intertwined with and indistinguishable from the action of the gun. Seeing and striking quite literally coalesce into a single cathartic expression, disclosing the possibility that it is a certain mode of visual apprehension as much as the rifle that has killed this creature.

Only a few years later, Henri Bergson would formalize this relation, arguing in *Matter and Memory* that the perception of objects is largely determined by an impending action.[12] As the eye impregnates the present with this virtual future, a world is brought into being around a latent motor event (the strike). We see, says Bergson, in order to act and in so doing configure a world around this impending action. Critical to this model is the fact that the perceiving subject is not the origin of this world. Rather, it is the entanglement of this "center of indetermination" with the visual field and its array of "extended objects" that bring into being new worlds. In navigating this field, the subject essentially subtracts from the object those qualities that conform to its intended use ("the size and color of a woodcock" in the case of James's hunter) and reciprocally fails to apprehend the latent or unrealized possibilities that are irrelevant to this instrumental disposition (those qualities that would identify the bird as a thrush rather than a woodcock). In the instance of automatic recognition and habitual response, the past is no longer presented to consciousness, despite its having partially structured the action. Instead, the world is experienced as a kind of pure futural medium in which seeing collapses into action.

It is this interpenetration that the inorganic agencies of drone violence seek to reproduce. In fact, one might even say that the function of the kill chain is to escalate the conflation of seeing and striking that plagues James's hunter. This prospect would seem to fly in the face of the drone program's obsessive intelligence gathering. (It is estimated that the average drone collects roughly twenty thousand hours of footage each month. The Autonomous Real-Time Ground Ubiquitous Surveillance System [ARGUS] discussed in chapter 2 of this book is capable of capturing eight years' worth of video in one day.)[13] In light of the extraordinary amount of information collected by martial networks, one must wonder how it is that the simple possession of a SIM card or traveling in a group at night can transform a subject into a viable target. Would not, in other words, this extensive archive serve precisely to safeguard against such simplistic, knee-jerk reactions? The sheer frequency of attacks like the one in Datta Khel suggests that signature strikes not only reject such a logic, but in fact operate upon a reversal of its basic premise. In the case of drone power, the intelligence collected by these networks is directly rather than inversely proportional to the deployment of nonspecific violence. It is in fact precisely the contradiction between the insatiable drive to accumulate information and the nonspecific nature of drone violence that allows the signature strike to duplicate these relations of targeting by inorganic means. Put simply, the more drones know, the less beholden they are to a specific, identifiable target. This counterintuitive relation is at the center of the logic of simple triggers, which appropriates the Bergsonian action-image in order to present a world in which the strike is, in effect, always already present.

On February 21, 2010, a Predator drone was called in to track three unauthorized vehicles traveling down a dirt road in Khud in the Oruzgan province of Afghanistan. After reviewing video of the passengers, the image analysts identified "cylindrical objects," which were quickly recast as rifles. One analyst noted cold spots on the subject's chests and soon became convinced they were concealed firearms. Before long the group's movements began to appear as "tactical maneuvers." Derek Gregory summarizes the systematic translations that the kill chain began to manifest: "Two SUVs and a pick-up truck became a 'convoy', cylindrical objects 'rifles', adolescents 'military-aged males' and praying a Taliban signifier."[14] These calculations culminated in an attack on the group of nearly two dozen individuals who, it was immediately discovered, were "shopkeepers going for supplies, students returning to school, people seeking medical treatment and families with children off to visit relatives."[15] As this translation of objects and events into strikes relies upon an automation of the symbolic realm, the technological might of the network is disproportionately steered toward the field of virtuality (the building and commandeering of conditions through reconnaissance) relative to the processes of actualization (the validation and implementation of the strike). This creates a strategic discrepancy between the profusion of data mobilized around the attack and the simplified relations of causality that structure the experience of the operators and analysts.

This bottlenecking effect transforms a vast and illegible archive into a coherent operational environment by compressing contingency into simplified "if . . . then" formations. While the strike itself is the most visible effect of this relation, this process begins at a primary computational level. It originates in the ongoing atomization of data that occurs at the hands of queries and/or metadata creation and is reiterated by an internal movement by which assets are rendered "true" via the corroboration of new data and/or readers. Graham Harwood describes this evolution in the following terms: "A low-ranking larynx fixed by GPS markers that connect by radio wave. The report cascades up a command hierarchy until the SigAct is 'true.' Then de-constructed into data-atoms, logical machines compress the contingency of the moment as another higher-ranking larynx calls across another radio wave."[16] After unearthing a command line MySQL search for "smile," Harwood speculates about how this fragment may have functioned within the kill chain: "[H]ow many children smile with open arms when a solider enters a village can be correlated with the key 'hostile' or 'friendly' at a particular grid reference. Does such calculation lead to a drone strike or a gift of chewing gum or chocolate?"[17]

The crude determinism of simple triggers is made possible by a complementary logic of tactical animism, in which the martial network's automation of the symbolic field produces realities that read as the self-revelation of the objects, events, and agencies it encounters. In this arrangement, smiles, group formations, and "cylindrical objects" serve not so much as the subject of the drone's eye, but rather as the origin of worlds organized around an impending action. This process relies on a kind of perverse object-oriented ontology that prompts the material realm to strategically jettison its passive status in order to materialize the strike as latent presence.[18] Jordan Crandall describes this relation in the following terms:

> As intelligence migrates into unlikely, shared sources, even those spatial and atmospheric, and agency is understood to be distributed and embodied in all manner of organic and inorganic actors, a concept of skill emerges whose source is in negotiation rather than domination: an alliance with material actors rather than an assertion of command over them. Here an actor works with a material rather than against it, cultivating an existing, emergent meaning rather than externally imposing one—a "knowingness" that is not simply categorical but affective and rhythmic. It transforms objects into situations, their contours not determined in advance but arising within the terms of the encounter.[19]

The mobilization of data that makes the signature strike possible disconnects intelligence from its humanist base, so that rather than offering definitive interpretations of observable phenomena, it prompts the material world to disclose a plurality of discrete realities that drones and their operators process as action-images. In this way, martial networks detach drone vision from both the realm of representation and the founding questions of surveillance: Who is this? What are they doing? While stopping short of automating the attack itself, these networks nonetheless utilize these dynamics to ensure that the reality in which operators engage their subjects is built around a specific action-image. The strike then becomes a simple matter of acting on the triggers that the network has deposited within the sphere of objects, events, and places. Faced with the world's autogenic presentation of itself as target, the delivery of a Hellfire missile appears as a given, a means of actualizing a world that already presents itself as true.

By integrating these complementary logics of simple triggers and tactical animism into its mode of presentation, drone art divulges a pervasive "hunger" that both structures the interfaces of drone operation and catalyzes the enfolded agencies that underlie its power. Exemplary of this dynamic are Trevor Paglen's short video *Drone Vision* and Joy Garnett's *Predator* paintings, both of which mobilize the visual iconography of automated war in order to present a mute and expectant world that can only be given meaning by way of the strike.

POSTAUTOMATION AND THE HUNGER OF DRONES

In the grainy videos of automated war that have entered the public sphere in the last several decades, the missile acts as a particularly important narrative device. Like the internal spectator of painting or the voice-over in film, it promises to lead the eye to a point of interest within the otherwise undifferentiated spaces of an elsewhere war, to assemble ungrounded images into a discernible world. This function is so well-established that when the appropriated drone footage of Trevor Paglen's *Drone Vision* (see figure 1) conjures this familiar visual language with staticky black-and-white images and ominous crosshairs, its seemingly banal content of empty landscapes and open skies is immediately overpowered by the sense of an impending strike. In fact, the interchangeable cloudscapes and nonspecific desert topographies that populate this appropriated drone footage seem to arrange themselves into a familiar non-place that awaits the arrival of a missile in order to become fully legible as place. The affective presence this latent violence produces cannot be described in terms of simple expectation, nor can the function of this impending attack be adequately conveyed in narrative terms, i.e., as a structuring absence. Not only is the Hellfire missile the sole agent capable of bestowing meaning on the metadata that accumulate at the margins of the image, but so is it the only way to bring this image into focus, so to speak. While it is commonplace to speak of events needing images in order to come into being, here it is the reverse: the image needs the event if it is to lay claim to a functioning diegesis.

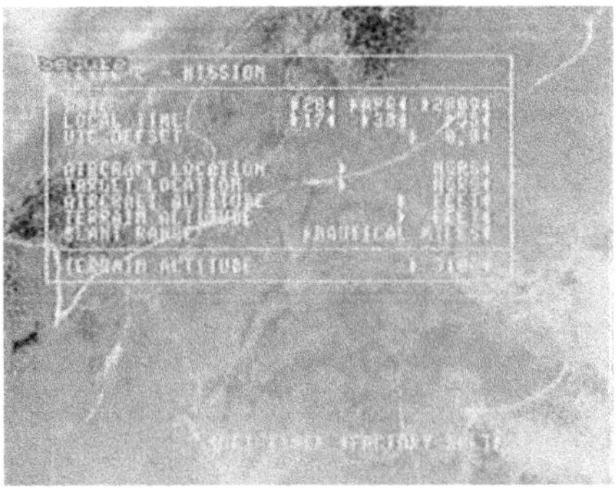

FIGURE 1. Trevor Paglen, *Drone Vision*, 2010. Video still. 5 min., 20 sec. Courtesy of the artist; Metro Pictures, New York; Altman Siegel, San Francisco.

And yet despite the unsettling sense of anticipation that builds with each replay of this extended video loop, the attack never happens in this video. Targets do not materialize. Narratives do not emerge. Having been disconnected from the networked relations of the kill chain by the artist's act of appropriation (the footage was originally secured by an amateur hacker), the unfulfilled desires of the drone are left to echo and intensify within the frame as an insatiable lack. The experience this creates for the viewer is not unlike the artist Graham Harwood's own encounter with the decontextualized data of the *Afghan War Diary*: "You soon find yourself becoming hungry for a device that will sort, join, and create new views out of the 91,000 comma-separated lines. It's as if the image of the parent machine emerges from the potential of the information in front of you. The text blobs scream out to be hung on indexes, laid out on the latitude and longitude of maps, and ordered across the epochs of time."[20] By removing the UAV from its native environment of the network, Paglen provokes the drone's eye to divulge a similar "hunger." Wandering through environments without an explicit focus or destination, this desiring gaze is oddly subservient to the world it encounters. It marks a specific type of attentiveness, a readiness to translate the generative flow of signifieds that emanate from the world below into violence. It is this "desire without an object" that penetrates the picturesque clouds and empty landscapes of this footage and manifests as anticipatory anxiety in the bodies these images leave unrepresented.[21]

The desire that *Drone Vision* discloses is driven by a regime of temporal condensations that operates across both technological and historical registers. As its simulations of the future are fleeting and fluid, made possible by the momentary alignment of real-time data, drone operation fuels a specific mode of urgency that Samuel Weber argues is represented in popular discourse by the phrase "target of opportunity." This phrase was used in the media to describe the alleged hiding place of Saddam Hussein and quickly became representative of a larger logic of the war on terror. As Weber describes, while the enemy had to be "identified and localized, named and depicted" in familiar fashion, what is new in the post-9/11 context is the "mobility, indeterminate structure, and unpredictability of the spatio-temporal medium in which targets had to be sited."[22] The very singularity of phenomena becomes intertwined with what Weber describes as "the generality of an established order, scheme, organization or plan, in respect to which the event defines itself as exceptional ["SigActs"] or extraordinary."[23] Clearly, responsiveness has always been rewarded in warfare. However, in the case of the signature strike, it is not so much that a given event presents an opportunity, but rather that the opportunity is the genesis of the event itself. As a result, the "opportunity" to strike is incessant, even eternal. This unique mode of urgency unifies the technical platforms of the kill chain, which in turn collectively produce a pervasive and insistent pull from without.

The capacity of modern drones to satiate this hunger relies on the elimination of the interval between seeing and striking, a possibility that was gradually realized in response to a series of incidents that took place during the hunt for Osama bin Laden at the end of the last century. In 1998, then President Bill Clinton ordered a cruise missile attack on a compound in which bin Laden was residing. In the time between the launch of cruise missiles and the striking of a target (generally, three to seven hours at the time), bin Laden was tipped off and escaped the compound before the missiles struck.[24] Two year later, a drone flying over an Al Qaeda training camp spotted a man that closely resembled the Taliban leader. Fearing collateral damage, Clinton famously refused to issue the order to strike, due to the presence of a swing set on the video feed. To create greater precision in targeting and eliminate the delay that had derailed earlier operations, a prototype for the Predator drone was unveiled in February 16, 2001.[25] For the first time, the surveillance apparatus was merged with the weapons delivery system, materializing a simultaneity of seeing and striking that would structure the modern kill chain.

As Paglen's piece visualizes the way in which these historical and technological relations manifest as the inorganic desire of contemporary martial networks, the cathartic images of the Gulf War (1990–

1991) missile camera become important interlocutors for the work. Despite plummeting viewers toward unsuspecting targets with a nihilistic pleasure, the bomb's-eye view of these familiar videos nonetheless retained a definitive and active subject position throughout its morbid trajectory. As a result, the nose cone camera issued a paradoxical mode of interpolation that organized the world even as it destroyed it. In comparison, Paglen's work dramatizes the ways in which the drone relinquishes this embodiment in favor of an externalized gaze in which the expectation of violence emanates from the material world itself as depersonalized percepts and affects. By prompting viewers to actualize the virtual presences embedded in these images of landscapes and skies at an imaginary level, the work reproduces both the relations of tactical animism that fuel the kill chain's relations of force and the enfolding of human agency that makes this violence possible.

Joy Garnett's *Predator* series similarly integrates the visual iconography of drone media into the frame (see figure 2). As these works reproduce screengrabs from drone footage in a painterly, abstract style, the objects, landscapes, and figures that populate the canvas come to reside in a liminal space between presence and absence. A faint set of targeting guides overlays what appears to be a lorry. Crude black rectangles pierce through swathes of gray to suggest rooftops. Vertical elements traipse through landscapes in order to present the bare minimum of a human form. Punctuating these scenes is the occasional vortex of brushstrokes, which conveys the explosion of a Hellfire missile. In pushing subjects to the brink of representability, these paintings draw a correlation between the aesthetic strategies of minimalism and the operations of drone power. If in his famous series of eleven lithographs, Picasso asked what is the fewest number of lines one can make to communicate "bull," then the drone's eye seems to ask what are the fewest indicators of a subject one needs to confirm their status as target. In these terms, the works' reduction of the subject to its bare essence unites with the logic of the *dividual*, forming a world that is endlessly divisible and open to reconfiguration. Garnett's paintings dramatize the temporal dimension of this endeavor as partial signifiers move between ground and figure, mirroring the process whereby data from disparate sources begin to align in formations of momentary fullness. This drama unsettles the scene to such an extent that the explosion eschews recognizable causality, meaning, or even location. In the place of the "north arrow," which would indicate "true north," the works feature a single letter "N," which appears strewn around the canvas, disconnected from any

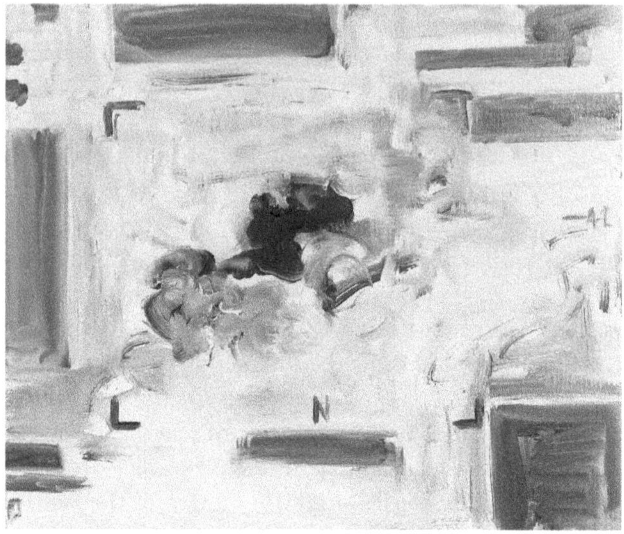

FIGURE 2. Joy Garnett, *Predator 2*, 2012. Oil on canvas. 30 × 35 inches (76.2 × 88.9 cm). Courtesy of the artist.

system of orientation that might successfully place the viewer. By performing the drone's ongoing eradication of orientation and ground, these works testify to what Derek Gregory describes as drone warfare's "undo[ing] . . . of the conventional ties between war and geography."[26]

Through this convergence of the canvas and the screen, the *Predator* series replaces the window, the dominant mode of understanding the canvas for most of the history of Western art, with the interface. Unlike the immersive and voyeuristic space of the window, this instrumentalized surface is wholly impassable. It not only resists investiture and identification but foregrounds its own mediating presence in the process. Johanna Drucker explains: "Interface is a dynamic space, a zone in which reading takes place. We do not look through it (in spite of the overwhelming force of the 'windows' metaphor) or past it. . . . The surface of the screen is not merely a portal for access to something that lies beyond or behind this display. . . . Interface, like any other component of computational systems, is an artifact of complex processes and protocols, a zone in which our behaviors and actions take place."[27] As these relations of the interface enter the painted image, the exchange between material and nonmaterial forces that typically defines the viewer's relation to the work is recalibrated. In the place of the "ontological cut," the break that divides the material sphere of the spectator from the virtual world of the image, the interface substitutes a deterministic continuity.[28] The diagrammatic signs that emanate from the dials, gauges, and sighting mechanisms that gather around the edges of Garnett's paintings elicit "input" from the user in order to complete the strike that is latent to these unformed brushstrokes. In this way, the "task-oriented and efficiency driven" nature of the interface transforms the fantastic projection of the spectator into an overdetermined relation of actualization.[29]

Vilém Flusser's understanding of the interplay between the subject and the "black box" of modern media is useful in unpacking the role of the spectator in Garnett's paintings and the position of the drone operator more generally. Housing a given set of "programs," the technical apparatus, according to Flusser, presents a delimited sphere of action, an image repertoire, which yields an illusory space of "play" for the user. As the device circumscribes a virtual field from which images, concepts, and events are actualized, this play does not typically expand beyond or challenge the finite parameters of the larger program. Rather than an extension of mental or physical capacities or a field of pure expressivity, the input of "functionaries" (Flusser's term for operators) serves primarily as the means by which the apparatus's preprogrammed possibilities are actualized. Signature strikes similarly delimit the sphere of actions, but do so by shifting the processes of automation from the sphere of action to that of experience. Their violence relies on relations of what might be called postautomation, a condition whereby the seeming free action of operators is enframed by an automated symbolic system in which such actions appear a priori. As Theodor Adorno's postwar diagnosis of the fascism of machines makes clear, these relations predate the modern drone:

> In the movements which machines demand from their operators, lies already that which is violent, crashing, propulsively unceasing in Fascist mistreatment. Not the least fault for the dying out of experience is due to the fact that things assume a form under the law of their purposiveness which restrict their interaction to mere application, without the surplus—were it that of freedom of behavior, were it that of the autonomy of the thing—which might survive as the kernel of experience, because it is not consumed by the moment of action.[30]

It is this founding loss of experience, the original loss posed by the industrial mode of production, that resurfaces as the precondition for the drone's unique mode of violence. However, as drones displace mechanical relations with networked digitality, their delimitation of experience is projected beyond the interaction of a user and an apparatus and into the world itself, where it is recoded as a series of simple programs. Thus, while the category of the user may be destabilized by the diffuse nature of the kill chain, it is precisely this "purposiveness" that drone vision extends to the human subjects it encounters. The result is that the very presence of the subject, made legible by the fulfillment of a set of conditions, becomes a purely means-oriented presence, a prompt to execute a world in which they appear as a target, in short, a simple trigger. Drones accomplish this by reversing the humanist bias of Adorno's condemnation, rendering the object world a source of agency that is not necessarily activated by human intervention but rather charged with creating situations in which subjects come into being. As viewers of Garnett's paintings configure the world in the frame around the attack, it is this operation they reproduce.

Through its extended loops of drone footage, Paglen's *Drone Vision* materializes the relations of drone power as a call from without, a pull that unsettles the directionality of violence. In similar terms, Garnett's paintings present collections of operative signifiers that require the strike in order to congeal into a coherent world. Together these works do not simply illustrate the logic of simple triggers within the frame, but rather issue a unique mode of address by which the agency of the viewer/functionary is enfolded into the inorganic agencies of the kill chain.[31] Though significant in and of themselves, the full import of these relations only becomes apparent when read in conjunction with the activation of the realm of objects that occurs under drone power. It is this latter scenario that is embodied by the work of Noor Behram and Nathaniel Hartman.

THE WORLD AS WEAPON: ACTIVATED (ART) OBJECTS

Since 2007, photojournalist Noor Behram has been photographing the devastation of drone attacks in the border region of Pakistan. Exhibited at Beaconsfield Gallery in London under the title *Gaming in Waziristan*, the images present the collapsed buildings, charred public spaces, and burned-out craters left by drone strikes. Alongside these vacant sites, the series also contains horrific images of the dead and severely injured, many of them children. Despite the affective intensity of these images of the deceased and the physical devastation caused by drones, perhaps the most intriguing of the series are the recurring scenes of fathers, husbands, and sometimes even children arranged before the camera, holding pieces of twisted metal, broken stone, and other debris in front of its lens for inspection. It doesn't take long to realize that in many cases these objects are shards of missiles and other components of a distant killing machine that has caused the scenes of devastation featured in the remaining images of the series.

On the most basic level, these catalogs of partial objects continue the operation of the series as a whole. They offer confirmation of the reality of the event, a rejoinder that cuts through the rhetoric of "surgical precision" and "targeted killings" like a knife. However, in their privileging of the object world, these photographs at the same time acknowledge a certain inadequacy of (human) narrative in this system of drone vision. In fact, the unconventional composition of these images can be read as an explicit rejection of the humanist biases of portraiture, as it reduces subjects to internal spectators whose personal narratives are subordinated to the object relations they display. On these grounds, Mark Dorrian has suggested that while these images oppose the distance of remote killing with intimacy and closeness, they are simultaneously bound up with a sense of remove from the event. According to Dorrian, this remove is the product of the belatedness that the medium of photography incurs at the hands of the accelerated present of the digital sphere. As such, he regards the work as a "violent cancellation of the possibility of witnessing [that] positions the photograph—which in its lateness, a lateness made absolute by death, is condemned to remain on the 'outside,' as it were—less as a vestige of an act of witnessing than as a record of its end."[32]

While Behram's images clearly wrestle with the representability of their subjects, the notion that this emphasis on the inorganic betrays a failure or inability to witness the event assumes a passivity of the objects on display. In fact, when read in relation to the processes of world-making outlined earlier in this chapter, the contradictions that Dorrian identifies seem less about photography's inability to keep pace with networked warfare than its active attempt to reflect back the conditions of drone vision. The spent casing of a Hellfire missile, the broken stones that were once a home, the craters that mark the point of impact—these signs serve as prompts to reverse engineer the logic of the kill chain, to draw out the series of conditions that produced the attack. The corresponding images of shoes, backpacks, and other personal belongings of those killed by drone attacks, neatly arranged before the camera as a personal shrine, comprise a similarly inverted system of simple triggers that conjure missing victims by interrogating the objects, behaviors, affiliations, and locations that conspired against them. Rather than being simple overtures to unrepresentability, these strategies allow Behram's images to reproduce the logic of conditions and parameters within the frame. In the process, the activation of objects that undergirds this system of violence comes to serve as the means by which missing subjects are symbolically reclaimed.

These relations of tactical animism also play a prominent role in Nathaniel Hartman's *ASM_frag* (2015), which presents a fragment of an exploded Hellfire missile that was smuggled out of Pakistan as evidence of a drone attack (see figure 3). The charred black surface and dramatic textures of the piece serve as indexical relations of a missing but wholly legible event. It is not simply the recurring media images of such strikes that allow the viewer to fill in the blanks, but also the likenesses that are inadvertently burned into the object's surface. Acknowledging the latter, Hartman describes the rough exterior of the work as "evok[ing] a sense of a desolate landscape [such that its] craters feel vast and [its] peaks [appear as] crumbling buildings."[33] Despite the undeniable affective currency of these physical traces, Hartman's sculptural form is in fact not the missile itself. Rather, it is a three-dimensional (3-D) printed replica produced from a photograph of the object that was widely circulated online. The work foregrounds the object's lack of provenance on multiple levels: the accompanying text openly explains its mode of production, the title mimics a filename so as to drive home its computational origins, and even the surface shows the presence of PLA filament, commonly used in 3-D printing. Despite this transparency, the experience of this ballistic shard nonetheless provokes the kind of affective charge typically reserved for an original. In this, *ASM_frag* turns upon a similarly paradoxical logic as that of Behram's images, in that its lack of "firstness" not only fails to detract from its claims to realism, but in fact seems to provide the basis for a certain kind of authenticity.

At the root of this relation is the fragment's dual status as document of a distant tragedy and artifact of a medial relation, a condition succinctly represented by Hartman's shorthand for the evolution of the piece: "Physical missile, to exploded fragment, to virtual fragment, to exploded physical form once more."[34] The telegraphic quality of these remarks betrays the rapidity with which this ontological movement collapses form and content, the capacity of its modulated realities to pull in opposing directions even as they testify to the certainty of their being.

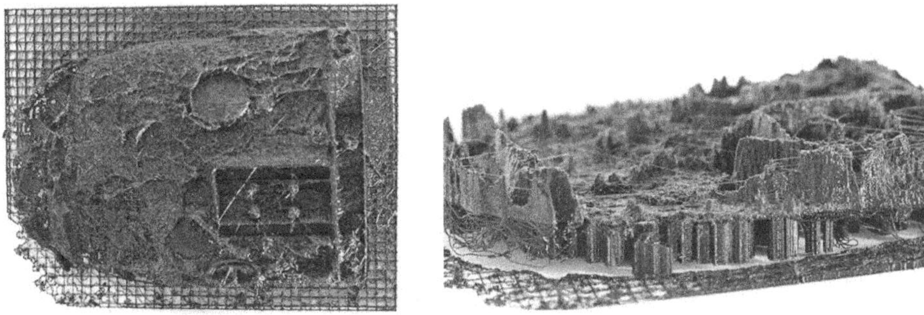

FIGURE 3. Nathaniel Hartman, *ASM_frag*, 2015. PLA 3-D print. 9 × 6 × 5 inches (22.9 × 15.2 × 12.7 cm). Courtesy of the artist.

From this perspective it becomes clear that as *ASM_frag* reproduces the strike across this genealogy of forms, a correlation is made between the becoming that undergirds the piece's mode of production and the violence it references on a symbolic level. The medium of 3-D printing plays a critical role in this relationship. Describing the process of fabrication, the artist states: "The structure had to be supported during the 3-d print, layer by layer, which created, strings, webs, and imperfections straying from the original 3-d model."[35] With the network's powers of deformation written into its material surface, the fragment comes to testify to a missing past event while at the same time disclosing its status as the condition of a future world. It is precisely these relations of immanence that, as an anonymous source from the US State Department put it, allows "three guys doing jumping jacks" to become a terrorist training camp.[36] One might even go so far as to say that this strategic movement through recognizable forms is the very prerequisite for an everywhere, everywhen war. The ease with which state violence transcends judicial and political barriers in the global counterinsurgency is after all inseparable from its ability to strategically elude and/or compromise enduring categories of conflict. As a result, the decomposing surfaces and intermedial origins of *ASM_frag* expand its succinct map of becoming beyond the confines of the particular object, articulating a more expansive movement between states that might best be represented as civilian to enemy combatant, bus stop to battlefield, targeted strike to endless war.

The logic of tactical animism that these works by Behram and Harman engage is made possible in part by the displacement of enduring relations of ocular-centrism with a pervasive modality of data-centrism.

The ramifications of this shift are perhaps most immediately apparent in the timeless now of the on-demand temporality of the database, which is integral to the NSS's (National Security Strategy) pursuit of a "war without duration" based in "continuous action" rather than "sequential" campaigns.[37] However, this relationship is unimaginable without the intervention of the more formative processes by which data are constituted as such. Bolstered by an enduring mythology of data as neutral, untouched, and "raw," acts of asset creation perform a kind of originary violence that the database amplifies and modulates around parameters of implied risk. As these trajectories of data production, access, and storage converge in the kill chain, simulated worlds are transformed into authoritative givens, providing the ground from which drone power extends beyond representation and into processes of world-making.

COMPUTATIONAL MEDIA, DATABASE WAR

In her introduction to the anthology *"Raw Data" Is an Oxymoron*, Lisa Gitelman points out that in order for data to function as such they must break with the immediacy of embodied experience. As that which is "imagined and enunciated against the seamlessness of phenomena," data, according to Gitelman, are constituted rather than simply collected and are therefore always already an interpretation.[38] Despite this inherent constructedness, data command a persistent expectation of objectivity, which in self-validating fashion conceals the interventions upon which this status relies. These relations prove critical to operations of the kill chain, where data-driven worlds preserve their authority by naturalizing a host of imprecise and/or motivated practices. These include the manual processes by which social and cultural reports (interviews with locals, meetings with politicians, etc.) are drafted in the field; the automated components that read and distribute this information after it is uploaded via the Secret Internet Protocol Router Network (SIPRNet); the structured queries that produce the search results presented to analysts, operators, and legal counsel; and the categories of criminality, gender, and ethnicity that organize these exchanges.

These interventions are concealed not only by the semiotic coding of this information as "data," but also by the "intuitive" interfaces that oversee the circulation of these assets within the kill chain. By prioritizing efficient interaction and totalizing representations, these intermediaries implicitly render "the work of inbuilt categories and representations invisible."[39] Exemplary of this dynamic is IBM's *Analyst's Notebook* software, which is used by drone analysts to "identify, predict, prevent and disrupt criminal, terrorist and fraudulent activities."[40] Weaving together data from over six hundred sources to form what the company touts as a "single cohesive intelligence picture," the program's visualizations sacrifice the complexity of context to the illusory promise of an exhaustive "world picture."[41] By repeatedly presenting these partial representations as complete worlds, such interfaces form a path dependency that not only prompts the underlying interventions of data to disappear at an operational level, but progressively eradicates entire fields of information in the process.[42] Clearly these remarks only scratch the surface of a much larger narrative concerning the supposed neutrality of data and their role in legitimizing the operations of the kill chain. While a comprehensive analysis of these interventions exceeds the scope of the current discussion (and perhaps the legal barriers posed by classified information), it is nonetheless worthwhile to consider in more detail two platforms that are foundational to nearly all of these instances, relational and postrelational databases.

Unlike earlier databases, which utilized "indexes, chains, and pointers" to structure information, the relational database, as Rieder describes, is comprised of "regular, uniform, and only loosely connected objects [that] can be ordered in a potentially unlimited number of ways at the time of retrieval."[43] On this basis, it was touted as a flexible, horizontal organization of information that would elude the cumbersome hierarchies of its predecessors and ultimately "free" the information it contained. However, like so much of the cyber-utopianism of decades past, these sentiments have been walked back dramatically in recent years as theories of modulated control have proven that power relations are perfectly capable of coexisting with immanence and becoming. This scenario is exemplified by the GeoCell database, a compendium of cellphone records used to target insurgents via drone strikes. Summarizing the basic operation of this database, Gareth Porter explains: "The phone numbers and call histories from . . . phones go into the database which is used to 'map the networks' [of social network analysis systems]. . . . If the phone records show multiple links to numbers already on the 'kill/capture' list, the individual is likely to be added to the list."[44] The data for this archive are largely collected from US-ordered shakedowns in Afghani prisons, in which guards confiscate cell phones, which are then turned over to the US military. As such, the data are already circumscribed by categories of criminality that at least partially determine the content of these communications. What is perhaps even more striking is the way in which information that would seem integral to assessing a subject's level of threat is curiously omitted by the organization of the database. Can the data definitively prove that another person was not in possession of the phone at the time when certain calls were made? Are there witnesses who can confirm that the subject in question was in possession of the phone during a particular time? How does this system account for the fact that many people living in these areas have the numbers of Taliban commanders in the contacts of their cell phones as a "survival mechanism?"[45] These would seem to be vital questions to ask in order to fully contextualize the data accrued by "link analysis," yet the organization and syntax of the GeoCell database ensures that they are not addressed in a structural or concerted manner. Commenting on the larger effects of these blind spots, Jeremy Scahill and Glenn Greenwald explain:

> Once the bomb lands or a night raid happens, you know that phone is there. But we don't know who's behind it, who's holding it. It's of course assumed that the phone belongs to a human being who is nefarious and considered an "unlawful enemy combatant." This is where it gets very shady. . . . According to [a] former drone operator, the geolocation cells at the NSA that run the tracking program—known as Geo Cell—sometimes facilitate strikes without knowing whether the individual in possession of a tracked cell phone or SIM card is in fact the intended target of the strike.[46]

As the elisions enacted by data disappear in the interval between query and results, the database comes to operate as a kind of call and response, an echo that reifies and legitimizes assumptions in automated fashion. In this way, the fluid and nonhierarchical relations of its relational structure quickly ossify into their opposites at the point of retrieval, articulating modulated hierarchies within relations of becoming.

In order to recognize and respond to constantly changing conditions, the kill chain relies on the nearly limitless ability of the database to accumulate and organize new information. Unlike the narrative structures of chronology-based intelligence, in which the introduction of new elements prompts the data to "tell a different story," relational databases are configured so that additions can be made without interrupting the continuity or experience of information. On these grounds, Katherine Hayles argues that while databases work through inclusivity, narratives rely on exclusivity. This inclusivity is, however, contingent upon a kind of absolute conformity to the self-describing nature of the system, which does not require that the user "go outside of the database to see what it contains."[47] Not only is the structure of this archive contained within the database itself, but so is the act of retrieval also bound by the governing logic by which queries take place. As a result, while narrative "always contains more than indicated by the table of contents," databases cannot process "indeterminate value," which materializes as a null value or not all.[48] Translating these observations to the GeoCell archive illustrates how the relational structure of the database allows the circumstances surrounding the SIM card to be strategically nullified. One might even go so far as to say that it is the very inclusivity and nonhierarchical relations of the platform that provide the basis for this exclusion.

Through these relations of erasure and nullification, the database achieves agency in the kill chain. The blind spots that emerge from these relations not only destabilize the category of the enemy combatant but also actively steer the apparatus toward civilian targets. Greg Porter explains:

> [T]he link analysis methodology employed by intelligence analysis is incapable of qualitative distinctions among relationships depicted on their maps of links among "nodes." It operates exclusively on quantitative data—in [the case of the GeoCell Database], the number of phone calls to or visits made to an existing JPEL (Joint Prioritized Effects List) target or to other numbers in touch with that target. The inevitable result is that more numbers of phones held by civilian noncombatants show up on the charts of insurgent networks.[49]

While Porter explains this tendency of signal intelligence (SIGINT) to integrate civilians into martial networks in terms of a general privileging of quantity over quality within the architecture of the database, these same processes can also be fueled by a scarcity of information. In fact, there are several known instances when targets have been struck based on a single social media post.[50] Clearly the persistent logical "fallacies" of drone power are not specific to the "link analysis" methodology, but also inform "patterns of life" calculations. To find confirmation of this one need only consider the case of journalist Ahmad Muaffaq Zaidan, who was placed on the National Security Agency's (NSA's) SKYNET database by virtue of traveling a specified route on a regular basis and meeting with known terrorists. Taken in the context of the kill chain as a whole, the inability of this system to distinguish his journalistic actions from that of an actual terrorist no longer appears as a momentary glitch or lapse, but rather a product of certain structural relations of the system.[51]

The only reasonable conclusion to draw is that civilian casualties are inevitable rather than accidental in the logic of the signature strike. However, as such an admission does not prove politically viable, the state must either justify such tactics as a necessary evil or deny their existence. With the latter approach made increasingly untenable by the public nature of drone strikes, attacks are often unwillingly acknowledged. Operating under the cover of "the state of emergency," the ensuing discourse posits this intervention as a last resort taken to preserve the safety of citizens and in so doing foregrounds the threat as the catalyst and justification for the suspension of the law. In the post-9/11 era, the longevity of such a state is correlated to the lack of a formal declaration. Just as the endless war requires a nonwar in which enemies are unidentifiable and conflicts are nonspecific, so too does its corresponding state of emergency necessitate the flexibility of a generalized condition, a militarized everyday that prompts the exception to silently become the rule. By mobilizing the contradictions of data and the architectures and agencies by which they are stored and disseminated, the relational database internalizes the dialectic between threat and aggression that fuels such a condition. As a result, it comes to provide the ideal medium for the immediate elevation of events to the level of threat and an equally immediate retaliation by force. However, this scenario can only be realized by way of a second-order operation in which the parameters and categories by which threats are produced are themselves subjected to the powers of immanence that inhabit the archive.

As databases such as the GeoCell are increasingly linked with other databases such as the Terrorist Identity Data Environment (TIDE) and Terrorist Screening Database (TSD), their assets are subjected to a joint constellation of postrelational (NoSQL) systems and genetic algorithms, which together constitute the computational ground from which targets emerge. In postrelational configurations, the categories by which threats are processed are no longer given in advance, and data are often left unstructured and varied. Unlike their relational counterparts, these classification systems are easily reconfigured in response to changes in circumstances and/or the integration of new data. As a result even the most basic category of drone targeting, the military-aged male, is opened to revision, so much so that even children can qualify as MAMs.[52] As Joseph Pugliese explains: "[C]hildren become 'dangerous' adolescents who then morph into military-aged males who, regardless of the fact that their identities are not known, finally become 'legitimate' drone targets."[53]

The engines of this postrelational logic are genetic algorithms that animate the database architectures from the bottom up in order to optimize the processing of large troves of data. Moving from solution to problem, these mechanisms mimic processes of biological evolution in their search for optimal combinations and global fitness rather than definitive solutions. In the context of the kill chain, this means that algorithms work to find or rather produce a possible identity to match a kill, rather than the reverse. Through this dual apparatus of genetic algorithms and postrelational architectures, the blind spots produced by individual iterations of the database are modulated around specific instances and retroactively distributed across existing fields of data. As a result, the parameters of the dialectic of threat and response spill over beyond the material and the existential, forming a field of immanence that amplifies both the immediacy of the former and the necessity of the latter.

In the face of these ultraresponsive worlds of impending violence, the tendency of drones to error on the side of excessive force can only appear as a compensatory gesture, a counterbalance to the imminent danger summoned by this dynamic archive. In this way, the martial database produces risk, which in circular fashion justifies the boundlessness of drone violence. The ramifications of this operation extend well beyond the immediacy of targeting. After all, it is not simply that the solidity of the category of the "enemy combatant" is undercut by these processes, but rather that the general population is itself inscribed within an unsustainable contradiction, serving as "both the source of dangerousness and [that which is] in need of protection."[54] Utilizing data from the *Afghan War Diary*, the *Endless War* installation by artists YOHA (Graham Harwood and Matsuko Yokokoji) and Matthew Fuller seeks to co-opt these operations of the martial database. By recuperating the tensions between visual and nonvisual data, real and virtual space that reside within this archive, the work attempts to disentangle the kill chain's relations of immanence from its program of violence. Central to this operation is the work's active sabotage of the coherence of representation and the linearity of narrative, a gesture that allows *Endless War* to engage with the war body on a presymbolic level.

WORLD-MAKING AND DATABASE: THE *ENDLESS WAR* INSTALLATION

Endless War presents the real-time processing of the more than seventy thousand documents of the *Afghan War Diary* on three large-scale

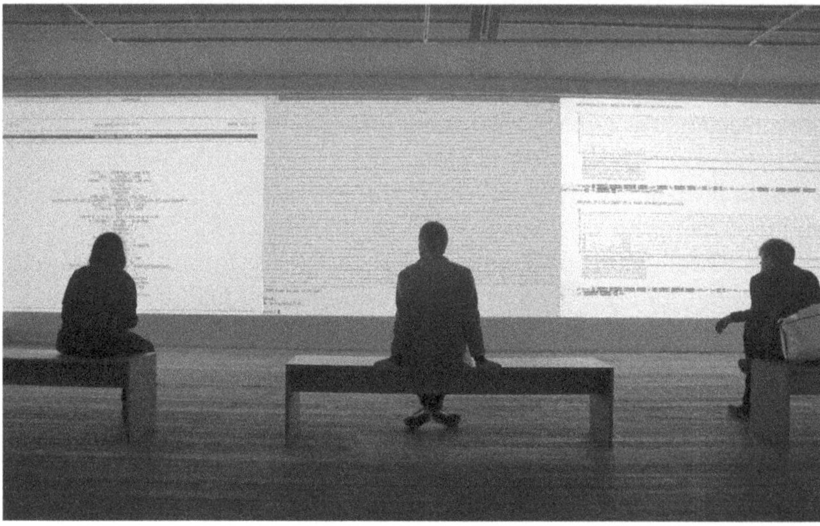

FIGURE 4. Yoha with Matthew Fuller, *Endless War* installation, 2012. Courtesy of the artist and Kunsthal Aarhus.

screens (see figure 4). Using N-gram analysis to sort and display these artifacts according to "repeated patterns of events, spatial information, kinds of actors, timings and other factors," the work exceeds the journalistic function of simple disclosure.[55] In fact, a critical component of the installation's intervention involves actively overriding the discrete metadata of these assets, which in their original state were "indexed spatially by the sixteen orbiting satellites of the Global Positioning System and temporally by the coordinates of atomic clocks."[56] This is made possible not only by the gesture of appropriation, which places the theater of operations in conflict with the here and now of the gallery space, but also by the work's sorting algorithm, which effectively unfreezes the history of these assets so that they might recirculate as potentiality once again.

This process of reanimation occurs on both visual and sonic registers. As a computer generates the sequences on display, the sound of its inner workings is picked up by a microphone and fed into the gallery space. The mechanical scrapes and electronic hums that accompany the oversized projections communicate the sense of work being performed from an inaccessible elsewhere, foregrounding the overall constructedness of this presentation in Brechtian fashion. On account of these relations, *Endless War* presents an irreconcilable tension. On the one hand, the work draws its power from the viewer's realization that the trove of information on display is the very raw material of war, the same data that intelligence analysts used to construct "charts, graphs, tables, and maps [that] conduct predictive analysis based on statistical trends."[57] On the other hand, the experience of these artifacts undercuts not only the certainty and coherence of such models, but also the larger expectation that the work will faithfully deliver "the war" in its entirety for the viewer. While this conflict is partially attributable to the overwhelming quantity and layered simultaneity that structures this particular presentation of data, it also draws from a more primary tension between computational processing and visual representation.

The question of how to represent the action of algorithms outside of the language of programming has perplexed computer scientists since at least the 1960s, when an impromptu debate took place between programmers in the pages of a leading journal.[58] After several individuals wrote in to express their confusion regarding the distinction between an algorithm and a mathematical formula, programmer Donald Knuth attempted to resolve the issue by claiming that an algorithm was not so much a formula as an "abstract method" for achieving a desired result. Critical to this distinction is the fact that a formula is a "finite" set of commands, while an algorithm can be separated from its mechanism. However, Knuth eventually conceded that this very same status of the algorithm makes it susceptible to irresolvable problems of representation: "I am forced to admit that I don't know any way to define any particular algorithm except in a programming language."[59] While the flowchart would endure as a functional diagram, it would achieve little cultural valence outside of the world of programming, leading to the kind of "public relations" problems that Christian Sandvig sees as plaguing the algorithm today.[60] The difficulties in representing algorithms can be boiled down to two primary obstacles. First, algorithms are defined by actions and as such elude a kind of essentialist definition or a "being in itself"; second, they are non-iconic, in that they do not have a visual presence that resembles what they are or what they do. These conflicts are a critical part of the experience of *Endless War*, which magnifies the algorithm's internalized tensions between representation and computation in order to disclose the relations of drone networks.

The visual excess of the work's three oversized screens contrasts starkly with the modest size of the text on display. As a result, reading these reports requires that the viewer approach an individual projection and maintain close proximity with it. In so doing, he or she is effectively disconnected from surrounding screens. From this position, the shifting images that occupy the peripheral view form an illegible flux that calls out to be read with an insistent urgency. In response, the viewer moves physically between projections, creating transversals between overlapping flows and frames of reference. As these spaces become interpenetrated with other possible realities, the most basic assumptions of the genre of installation are undermined. As Borys Groys claims: "[I]nstallation [typically] operates as a reversal of reproduction. [It] takes a copy out of an unmarked, open space of anonymous circulation and places it—if only temporarily—within a fixed, stable, closed context of the topologically well-defined 'here and now.'"[61] *Endless War* inverts this operation by applying the processes of circulation to the typically discrete "here and now" of the work. In so doing, it summons drone spatiality, which Derek Gregory describes as "transcending the geopolitical, connecting it to the biopolitical and the geo-economic, but also of tracking space as a 'doing', precarious, partially open and never complete."[62]

In negotiating this modulated space, one has the sense of committing a transgression, a feeling that is bolstered by the intimate proportions of the desktop writ large upon the gallery wall. However, as the "smoking gun" seems always to be on the virtual horizon, the installation elicits a vigilance on the part of the viewer. Indeed, it is as if its underlying algorithms egg the viewer on in pursuit of intelligibility. From the dense lingo and acronyms that fill the reports, one gathers that a task force was directed to a suspected IED site. Then another slide intervenes that describes a white flatbed truck with four bodies in the back. A flow of places, actors, and events surrounds the viewer, without congealing into legible totality. This dynamic flux pushes toward coherence, then pulls away, falling back into fragmentation just as it begins to articulate the possibility of another reality. The multiplicity of these worlds is not simply the product of fluid and unpredictable modes of force contained within the logic of an all-seeing imperial eye, but rather a spatiality that auto-reconfigures on demand, producing discernible places and identities by way of the fulfilment of parameters, conditions, and probabilities.

By turning the contents of this martial database inside out, the work materializes a unique multimodal address that summons a new war body. As indistinct sounds synchronize with shifting onscreen images, the symbolic processing of a transcendent and discerning ego is overtaken by the automatic, proprioceptive responses of the flesh. Sense certainty, semiotic coherence, and the mastery of narrative are undermined by the proliferation of non-iconic modes of information and the fragmented spatiality of the installation. As these processes suture the body into an ecology of machinic agencies, representational schema give way to prepersonal affective registers. The result is that one *feels* the presence of the war at a virtual, presymbolic level, as that which shadows the present but never fully materializes *as* present. The artists describe this experience in terms of a consolidation of the viewer into an "entire sensory and intellectual apparatus of the military body readied for battle. . . . As the ensemble of technical objects and flesh congeal, they create an organ to collectively act to rid itself of some perceived threat."[63]

The full significance of these relations becomes apparent when comparing the work to the combat film. As chapter 5 discusses in greater detail, the mainstays of the war film—its hyperindividuality, heroic codes of risk and sacrifice, and even the very idea of discrete and enduring goals—conflict with drone warfare, which seeks to eliminate the possibility of harm through asymmetric relations, replace proper nouns with situational valences, and instantiate an "everywhere war" that refuses temporal and spatial boundaries. While the combat film has struggled to translate its affective horizons of risk and heroism to the realm of automated war, *Endless War* offers something of an alternative by presenting a month-long processing event that reproduces rather than represents the dynamics of this new mode of violence. It does this by accelerating the tensions within the algorithm, prompting it to project artifacts, plans, and possibilities alongside the amplified audio of the machine that has rendered this shifting reality. Through this tension, the work returns these events to a moment before their inscription into processes of world-making and in this way manifests the raw force of martial networks rather than inscribes them within a definitive narrative.

And yet, as the visitor escapes the lure of the oversized screens and begins to explore the space of the installation, the virtual nature of these relations is qualified by an opposing trajectory. Upon leaving the light of the screen, the viewer discovers the computer hard drive, the origin of the disconcerting sounds and images that have comprised the work. The unique mode of presence articulated by the tower and mixing board as they sit upon their wooden stand conveys the stubborn persistence of materiality within this flux of becoming. Illuminated by the glow of the projections, this form, which teeters between virtual possibility and actual presence, seems to circumscribe a movement from which targets emerge. In so doing, it reminds one of the ability of these computational processes of the kill chain to manifest in and as the bodies, landscapes, and objects of the everywhere war.

Endless War discloses the computational origin of the hunger of the drone landscape. It does so by exceeding the visual and engaging instead with a more primary mode of becoming, which intertwines relations of data storage and access with bodily sensation. As the dialectic of threat and response occupies the flesh, the viewer effectively performs the collapse of "seeing" and targeting that follows from the militarized action-image of martial networks. The genre of installation proves critical to this possibility, as not only does its material and situated nature push this experience of drone power toward the corporeal, but so does its inherent intermediality provide fertile ground for reproducing the networked relations of the kill chain. The latter relationship is especially critical to Laura Poitras's *Bed Down Location* (2016), which self-consciously integrates this multimedia address of the genre into the experience of the work by way of the camera. As installation has historically operated as a convergence point for multiple media such as sculpture, architecture, painting, new media, and performance, it is only logical that it register the increasing presence of the mobile device in the museum more immediately than other forms of art. In the process of doing so, a unique mode of address is forged, one that negotiates the pervasive prompt to photograph with the promise of an embodied and immediate experience. Reading Laura Poitras's *Bed Down Location* through this lens will not only dramatize the larger consequences of tactical animism and simple triggers for subjects on the ground, but also lay the groundwork for the discussion of photography and drone power that takes place in the next chapter.

CONCLUSION: PERFORMING DRONE DEATH

Reigning narratives of twentieth-century art present the genre of installation in terms of the introduction of corporeal relations into the predominantly visual environment of the museum. Drawing from the interventions of Minimalism, this activation of the viewer as a phenomenal and fully embodied subject marks a pivotal interruption of modernism's temporal and spatial idealism, opening the door to familiar discourses of contemporary art such as site specificity, relational aesthetics, and even social practice. While there is certainly validity to this lineage, such a reading of installation can obscure a crucial duality at work within these sites. One can easily lose sight of the fact that the viewer's engagement with this work is quite often simultaneous with an alternative modality of interaction, which reiterates relations of passive spectatorship within a reflexive register. Put simply, the experience of installation also involves watching the interaction of others with the work, processing the images that are produced by these encounters, and reading the piece as a whole in terms of this second-order drama. The arrival of the mobile device and the subsequent easing of prohibitions against photography in museums have solidified this dual nature of installation, such that the dynamic exchange between the production of images and the immersion of embodied experience has become foundational to the genre. Exemplary of this relationship is Laura Poitras's *Bed Down Location*, which utilizes the camera's interjection of distance in order to fully disclose the embodied performances of drone death that the installation elicits.

Visitors to Poitras's *Astro Noise* exhibition (2016) at the Whitney Museum encountered an elevated pedestal on which they were invited to lie down and gaze up at a ceiling-mounted screen. Above them was projected video footage of the night sky of Yemen, Somalia, Pakistan, and other hot spots in the drone war. The banal tranquility of these slow, distant images is hardly interrupted by the occasional presence of a tiny drone, which passes across the field of view without incident. Along with the darkened room and overall scale of the screen, this viewing arrangement conjures a vaguely cinematic mode of spectatorship in which identification with the camera is synonymous with the rejection of the ambulatory body. And yet the overall experience could hardly be equated with the kind of submission to the image that Jean-Louis Baudry famously described in terms of being "chained, captured, [and] captivated" by the screen.[64] The intermittent presence of visitors passing through the space, the sounds of chatter coming from elsewhere in the exhibition, and the free movement of attention between the work and these quotidian dramas all undercut full immersion and instead prompt an investment in the screen as screen. This insistence upon "looking at" rather than "looking as" reiterates the distance with which those in the Global North regard drone violence. As such, it operates not as a modernist device that introduces criticality into the otherwise passive relations of the spectator, but rather as a means of rendering the visitor unknowingly complicit in the asymmetry of drone power.

While the pedestal is the host for the embodied relations the work engenders, it is also at the center of a complementary drama of representation. In the context of the museum, such device functions as a performative that both issues a directive to the subject and transforms the object it supports.[65] Put simply, in prompting the viewer to look in a codified way, the pedestal asserts the object's entry into the realm of art. In this regard, it is exemplary of Jacques Derrida's discussion of *parega*, those accessories to the work (the frame, curtains, pedestals) that appear as mere decorative flourishes but are in fact constitutive of the artwork's status as such.[66] In the case of *Bed Down Location*, this constitutive power is magnified by the size and overall anachronistic grandiosity of this raised platform, which insists that the visitor both participate in the installation *and* gaze from afar at the performance of interactivity it displays. Recognizing that such cues to look are increasingly indistinguishable from cues to photograph in contemporary culture, the work consciously integrates the camera in order to make visible a unique performance of drone power.

As visitors lie down on the pedestal to gaze up at the expansive screen, their bodies take on an extreme passivity.[67] Like lifeless matter caught in the otherworldly glow of the LCD screen, they appear as corpses, casualties of the vertical relations of viewing. There is a sense in which these bodies fashion themselves as such in order to enter into visual relations with the drone, that the body is always already a target within this system and is cast toward death from the moment it enters the martial networks of the drone's eye. In this perverse bargain, one encounters a revision of Bishop George Berkeley's dictum that "to be is to be perceived." In the world of the drone, to be perceived is simply to no longer be. The viewer's relation to the work dramatizes this nonreciprocal nature of drones, their transcendence of purely visual modes of power, and the automation by which their technological infrastructure transforms subjects into targets. At the hands of these forces, the bodies of visitors unknowingly perform drone death.

The camera does not simply document this performance but rather interjects an additional dimension by way of its relation to digital networks. As the repeated performances of this photographic situation circulate through media networks, a shifting archive is formed, an archive in which virtually the same scene reappears with different actors, origins, and meanings. Within this collection of images, not only do these performances of drone death take on a generic, unformed quality, but so too do the subjects of the camera lose their specificity in the process. Their visual presence comes to designate what Deleuze refers to as "a life," a form of immanence that is "subjectless, neutral, and preceding all individuation and stratification, is present in all things, and thus always immanent to itself . . . an immanent life carrying with it the events and singularities that are merely actualized in subjects and objects."[68] In the context of the installation, this generic body attests to the semiotic helplessness that comprises drone subjectivity: the young man's transformation into a "military-aged male," the reconstitution of a meeting of tribal elders as a clandestine operation, and the cylindrical object's transubstantiation into a weapon. It performs the condition that subjects of drone power live with on a daily basis—namely, that this moment might give way to another identity and another life that would nullify this one. However, one need not scan the Internet for photos from the work to find evidence of this relationship. All of this is confirmed for the visitors themselves as they encounter a live video feed elsewhere in the exhibition, which features infrared bodies of visitors currently lying beneath the projection. In this moment, it becomes clear that these faceless, genderless, and motionless forms are the placeholders from which the drone constructs its worlds.

In its focus on possible identities, meanings, and events, drone power enacts a shift from producing a definitive or complete world picture to overseeing the ground from which reality and its representations emerge. The mobilization of data that makes this project possible conflicts with the historical goals of surveillance and reconnaissance operations. Instead of identifying discrete individuals, the kill chain collects the partial traces of metadata in order to produce the proper nouns necessary for a strike. As this operation entails a larger rejection of the humanist base of media, drones both eschew identity at the level of content and refuse the notion of a singular viewing or sensing position at the site of reception. *Bed Down Location* renders these relations of drone targeting legible by drawing upon a convergence between the increasing presence of photography in the museum and the installation's insistence upon immediate embodied experience. While these tensions may be recouped within the confines of the museum, the larger prospect of photographing drones remains fraught with contradiction. Specifically, as drones resist stasis, continuity, the discrete subjectivity that inhabits the singular vantage point, and even the larger category of representation, the medium would appear ill-suited to capture its inner workings. In pursuing these conflicts, the following chapter considers how the drone's processes of world-making resurface in the context of the still image. Central to this discussion is the capacity of the still image to introduce these contradictions into the frame in such a way that provokes a dynamic symmetry between the civilian networks of social media and the martial networks of drone power.

CHAPTER 2

How to Photograph a Drone

The Nesting Logic of Vertical Empire

As objects, drones evoke equal parts fascination and unfamiliarity. Their shrunken proportions, absence of a cockpit, and undersized engines confound the expectations one has for a military aircraft even as the presence of state insignias and impressive payloads remove any doubt about their purpose. An unsettling groundlessness emanates from these ghostly bodies, which seem to resist the viewer at some primary level. Eluding the here and now of phenomenological presence, their matte surface, interchangeable silhouette, and overall sense of impenetrability divulge only the eternal elsewhere of drone power. Within the web of conflicting associations, histories, and sensations these objects articulate, one does, however, find a single familiar, even anachronistic feature: the camera. Conjuring collective anxieties of omnipresent surveillance systems that catalog the lives of their subjects with an unblinking attentiveness, the drone's oversized lens is oddly consoling, if for nothing else than the coherency of its message. Its subjects cannot help but imagine that behind this panoramic eye lies an exhaustive visual record of the life below, an archive whose verisimilitude would rival Borges's life-sized maps.[1] However, while this technology certainly participates in extensive surveillance operations, reading drone power as a simple escalation of everyday processes of optical surveillance fails to fully account for the rupture posed by martial networks. For this reason, the camera lens and its associations of an all-seeing eye are perhaps best understood as red herrings in decoding drone power. After all, it is not the map that this weaponized media ecology desires, but the landscape itself.

As discussed in the previous chapter, signature strikes shift the operation of targeting toward a prepersonal set of circumstances in which proper nouns exist as strategic possibilities. While threats are conceived of independently of identity in this system, the latter nonetheless resurfaces in the kill chain via processes of autogenesis in which the target essentially performs its status as such. In this way, drone networks serve to justify the killing of anonymous subjects as much as to locate known enemy combatants.[2] To sustain these operations, media maintain a tenuous relationship to representation at the level of both reconnaissance and strike. In terms of the former, partial images are conjoined with data to bring preexistent worlds into being. Rather than delivering static conditions, these images are interpenetrated by an immense archive of diagrammatic signs that includes heat signatures, bodily movements, chemical compositions, and GPS coordinates in order to build an instrumentalized diegesis around the attack. The result is a shift from a representational matrix toward what Mike Hill calls "fractal media," mathematical simulations that "rethink" the profilmic event rather than visually represent it.[3] A similar subordination of representation occurs at the strike. Here, the image is not so much a prosthetic extension of human agency as a frame that brings latent narratives into being by decoding embedded triggers. In this capacity, ocular media produce partial artifacts that achieve momentary fullness by way of the strike but are rarely definitive or authoritative in themselves. This undercutting of representation is part of a more generalized waning of ocular-centrism, which is exemplified by the role of video within martial networks.

A common misconception about drone warfare is that it takes place within a phenomenological unity whereby operators share the same time and virtual space of engagement as that of their targets. It is assumed, in other words, that drone pilots in Arizona respond to conditions that are happening in real time and that their actions produce events that are directly visible as they are executed. In reality, the "live" video feeds of drones are plagued by a debilitating lag, which effectively means that "all operators have to aim at is the slightly obsolete image of an earlier situation."[4] To exploit the latency of this system, counterinsurgents are trained to run in zigzag patterns when they are spotted by a drone. These practices bear witness not simply to the lag of drone video feeds, but to a larger subordination of the visual within martial networks, a relationship that exemplifies what Joseph Pugliese refers to as the "blind-seeing" of the UAV.[5] Here visual perception is no longer a matter of extending the eye via the circulation of light, as with photographic-based media, but of corroborating the information provided by sensors, informants, and satellites, after the fact.

While drone vision may maintain a parenthetical rather than prosthetic relationship to the agencies of the kill chain, visual representation is far from obsolete within this ecology of media. In fact, it is precisely this fragmented and subordinate status that allows drone vision to be conjoined with data of varying formats to bring fleeting totalities into being. The optical images that undergird these practices of "social network analysis" must relinquish singular vantage points and fixed temporalities. They must, like Behram's cathartic images, mirror back the conditions of machinic vision in order to ceaselessly become other. Consider, for example, the army's ARGUS system, which stitches together multiple video feeds from a single drone to produce an interactive map of a given neighborhood. From a live simulation of the city, users can zoom in to pedestrian traffic on a given street, follow a vehicle of interest, or even map the audio of intercepted telephone calls spatially onto this representation of the city via GPS coordinates. As these events are cross-referenced with previous data, they can also be read against historical patterns, ensuring that the extraordinary spatial cover of this technology is matched by equally limitless temporal parameters. Grégoire Chamayou describes the dynamic nature of this assemblage in the following terms: "Once [this] movie of every life and everything is completed, it could be rerun thousands of times, each time focusing on a different person, zooming in on him or her so as to reexamine that person's own particular history. One could select scenes, rewind, replay and fast forward, navigating as one wished through not only space but also time."[6] In this environment, the singular image will not suffice. Its articulation of a static and discrete world is at odds with the space-time manipulations of the global counterinsurgency. However, rather than disappear, it persists in partial forms. This status provides the conditions for what Harun Farocki calls "operative images," in that it allows drones to divide, categorize, and interrupt the visual field so as to reassemble rather than represent the world.[7] Thus, while it may be tempting to read the temporal lag of video and other interstices within drone vision as simple technological shortcomings, they in fact serve as critical entry points for the kill chain's processes of world-making.[8] Specifically, the belatedness of the visual image ensures not only that data-driven simulations are formed into authoritative givens before the image takes place, but that the ensuing representations are in turn structured by the impending violence of the strike.

These formative tensions between optical media and drone operation are especially prominent in the context of photography where the ability to break with foundational ideas such as resemblance, stillness, and authenticity serves as the condition of possibility for its entry into drone networks. In order to draw out the formative tensions that undergird drone photography, this chapter begins with a brief discussion of works by Fiona Banner, Nadav Assor, and Penelope Umbrico. Though representing a diverse set of media, these projects dramatize three critical points of conflict that shape photography's relationship to drones: the dynamic collision of existing symbolic systems with drone media (Banner), the constitutive movement that animates the image within martial networks (Assor), and the temporal interruptions posed by the eternal now of the kill chain (Umbrico). This discussion will then provide the foundation from which to approach the work of James Bridle and Trevor Paglen, which strategically accelerates these conflicts in order to reproduce drone power as both a system of internal reference within the image and a generalized movement across civilian digital networks. As the engagement of photography with the vertical empire of the drone becomes contingent upon a strategic breakdown of the stillness of the image, the function of the camera in this work is no longer to produce singular representations that bear witness to violence. Instead, it serves to circulate potential images, which both reproduce the relations of drone power and summon new worlds.

CAN THE IMAGE SURVIVE THE DRONE?

In Fiona Banner's short video, *Phantom* (2015), the camera follows printed publications as a powerful wind blows them across a parking lot. With each new sequence, pages begin to turn in an uncanny fashion, until these books eventually lift up and scurry across the concrete. As a familiar shadow soon appears in the frame, the viewer quickly realizes not only that the source of this gaze is a drone, but that the wind that animates these texts is produced by the blades of its rotors.[9] With this realization, a recurring drama soon becomes legible. At a distance, the content of these pages stabilizes as legible signs. Then, as the drone approaches and the wind intensifies, text and image are reduced to an indecipherable blur of flipping pages.[10] As these encounters associate the presence of the UAV with the interruption of a predigital symbolic system, the work performs the unraveling of reference that undergirds drone power. The repetition of this moment of symbolic rupture both clarifies the mutually exclusive relationship between these media and drives home the extraordinary violence it makes possible. Siobhan Leddy's description of the experience of the work succinctly summarizes this dynamic: "As *Phantom*'s drone draws closer to the book, its downward airflow whips at the pages. It becomes impossible to focus on a single point, as sheets are flipped and torn from the spine. At the very moment that we try to draw conclusions from what we see, what we see becomes something else entirely."[11] As this machinic mode of sensing clashes with the symbolic register of text and photograph, the drone silhouette articulates a nonsymbolic order that overwhelms and redistributes these systems. In this respect, *Phantom*'s repeated performances of nonintegration do not simply disclose the incompatibility of representational media such as photography with the drone's mode of vision, but rather push it to unsustainable thresholds so that the sign breaks down into illegibility. As this negation serves as the basis for an undetermined virtual field, the work dramatizes the ability of drone power to transform "what we see [into] something else" by way of the semiotic breakdown of text and image.

As part of a 2014 exhibition at the Julie M Gallery in Toronto entitled *Lessons on Leaving Your Body*, artist Nadav Assor displayed two video loops that showed the flight of a drone from conflicting vantage points.[12] In *Chariot*, the viewer sees the operator from the point of view of the UAV as it elevates into the sky, while *Ascender/Descender* shows a static long shot of the same figure in profile (see figure 5). The diptych format appears to present the controlling gaze of the operator as the agency by which these two disjointed vantage points are held together. Staring into the handheld monitor of the controller in both videos, this onscreen presence would seem to correlate these two conflicting spatial regimes, serving as both origin and object of a prosthetic gaze. And yet it soon becomes clear that this organizing agency is itself caught up in a unique form of movement that undercuts the possibility of such a reading.

As the first panel discloses a rapidly expanding territory below, the fixed parameters and static horizon of the adjacent long shot come undone, effectively telescoping the aerial shot through the earth. On account of this interpenetration, it is the landscape as much as or more than the drone itself that takes on the movement of flight. Most imme-

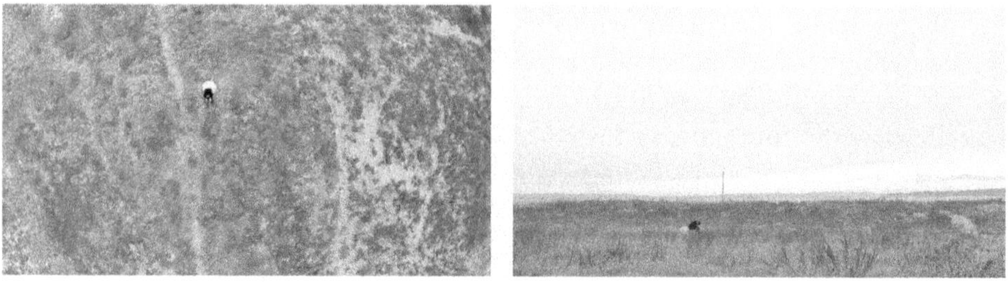

FIGURE 5. Nadav Assor, *Chariot* and *Ascender/Descender*, 2014. Stills. Video loops taken from *Lessons on Leaving Your Body* for exhibition at Julie M Gallery, Toronto. Courtesy of the artist.

diately, this movement interrupts continuity and eradicates the horizon, initiating a groundlessness by which the relative terms *ascending* and *descending* forfeit their direction. However, as the locomotion of the drone conflates movement within the frame with movement of and between the frame, it also subjects the objects on display to a process of becoming that displaces the singular referent with the multiplicity of martial networks.[13] In the process, the visual cues that would otherwise indicate authenticity—the supposedly neutral framing of the long shot, the uncut extended take—become containers for the drone's processes of world-making. While these works by Banner and Assor dramatize the productive conflict of drones with the spatial continuity of optical media, Penelope Umbrico's project, though not explicitly connected to the UAV, discloses the temporal foundation of these tensions.

In her ongoing series, *Suns (from Sunsets) from Flickr* (2006–), the artist mounts monumental photo installations based on a simple keyword search for "sunset." Using the results from Flickr, Umbrico crops the images closely around the sun and then uploads them to the Kodak website to make four-by-six-inch prints. These snapshots are then arranged in the gallery via a variety of formats, from expansive grids to vertical, totem-like structures. As an uncanny sense of sameness emerges between the photographs of users from around the globe, the work provokes a clash between the discrete "decisive moment" of photography and the sense of a singular expanded "now" of the networked database. The modular and interchangeable quality that the image maintains within these installations allows this slippage to infect the boundaries of the archive itself so that even the title of the work becomes unfixed:

541,795 Suns (from Sunsets) from Flickr (Partial) 01/23/06
2,303,057 Suns (from Sunsets) from Flickr (Partial) 09/25/07

3,221,717 Suns (from Sunsets) from Flickr (Partial) 03/31/08
5,911,253 Suns (from Sunsets) from Flickr (Partial) 08/03/09
7,626,056 Suns (from Sunsets) from Flickr (Partial) 07/17/10
8,730,221 Suns (from Sunsets) from Flickr (Partial) 02/20/11
10,291,373 Suns (from Sunsets) from Flickr (Partial) 01/12/12
13,806,070 Suns (from Sunsets) from Flickr (Partial) 11/01/13
18,297,350 Suns (from Sunsets) from Flickr (Partial) 04/16/14
21,314,840 Suns (from Sunsets) from Flickr (Partial) 05/14/14
27,709,969 Suns (from Sunsets) from Flickr (Partial) 05/05/15
27,694,473 Suns (from Sunsets) from Flickr (Partial) 05/11/15
30,240,577 Suns (from Sunsets) from Flickr (Partial) 03/04/16

As these spatially and temporally disparate suns coalesce into a single luminous source at the hands of the database, a totality emerges that is contingent upon its ability to break down and reconstitute itself around revised conditions or redefined parameters. This aggregating sun undulates and shifts as it teems with new entries, producing an eternal luminosity that quenches photography's insatiable desire for light with the intoxicating presentism of digital culture.

The work's disavowal of the intervals of night and day summons the same perpetual illumination that undergirds the casino's twenty-four-hour operation, the global stock market's perpetual present, the "just in time" modality of post-Fordism, and most recently, the global counterinsurgency of drone warfare.[14] This placeless, eternal now is conjured every time drones corral real-time data midflight in order to verify and produce targets. It reappears when "patterns of life" telescope past and future through the present in order to materialize a strike. On the other side of the globe, it provides the basis for drone labor as operators work across shifts to shadow targets for days without interruption, ensuring that surveillance functions as a condition or state rather than a singular event. In this way, drone operation personifies a larger displacement of the astral body of both agrarian and industrial modes of production by the on-demand temporality of late capitalism and its preferred mode of conflict. Natural light and its articulation of continuous space and linear streams of time conflict with the durationless character of the global counterinsurgency. As these works by Banner, Assor, and Umbrico confirm, drones fly by a different luminosity, and the photograph's claim to a discrete now is quickly overcome by this eternal sun.[15]

How to photograph the drone and its relation to its database sun? Put differently, how to sever the enduring relationship of photography to the astral sun? The knee-jerk reaction to this seemingly straightforward question would be to insist that one simply confront the relations of invisibility and secrecy that surround the drone by capturing its physical presence with the camera. However, the drone itself is only a representative of a displaced elsewhere from which its power emanates, a ruse to conceal its seemingly infinite tolerance for becoming behind outmoded concepts of presence and optical realism. As *Phantom* illustrates, attempts to "capture" the drone in photographic form inevitably reproduce this misrecognition by rendering the telepresence of the drone authentic and intersubjective. Put simply, drone "hunting" captures only the hunter.

In order to reproduce the kind of ontological movement featured in Assor's videos, the work of James Bridle and Trevor Paglen seeks to amplify rather than resolve the inherent tensions between the still image and drone power. In so doing, these projects free the medium of photography from its historical charge of freezing an instant so that it might become a vehicle for the unique movement of the kill chain. Bridle's *Drone Shadow* installations enact this reversal by presenting partial images of UAVs, large outdoor silhouettes that require photographic mediation to be made fully legible. As these interventions serve as a prompt to produce and circulate images, both the authority of the drone's presence and the witnessing function of the photograph break down. In the process, the image comes to take on a future orientation that articulates drone violence in terms of a shifting field of virtuality rather than static representation. While Bridle summons the conditions of the drone by displacing its presence across a field of future images, Paglen's *Untitled* series introduces the movement of the kill chain within the frame itself. These large-scale photographs of drones mobilize art historical references to the sublime in order to dramatize the operations of the "kill box," an algorithm used in drone operation that processes events in terms of spatial and temporal blocks. Through this union of martial networks and the conventions of landscape, Paglen's work initiates a movement between drones and the natural world that reiterates what Peter Sloterdijk calls "atmo-terrorism."[16]

Drone Shadows: Post-Internet Art and Partial Objects

In his *Drone Shadows* series, James Bridle draws chalk outlines of an MQ-1 on the street corners and parking lots of cities such as London and Istanbul (see figure 6). As if marking the scene of a crime, the aerial view of these monumental diagrams enacts a profound displacement whereby the forces of Western imperialism turn back upon themselves,

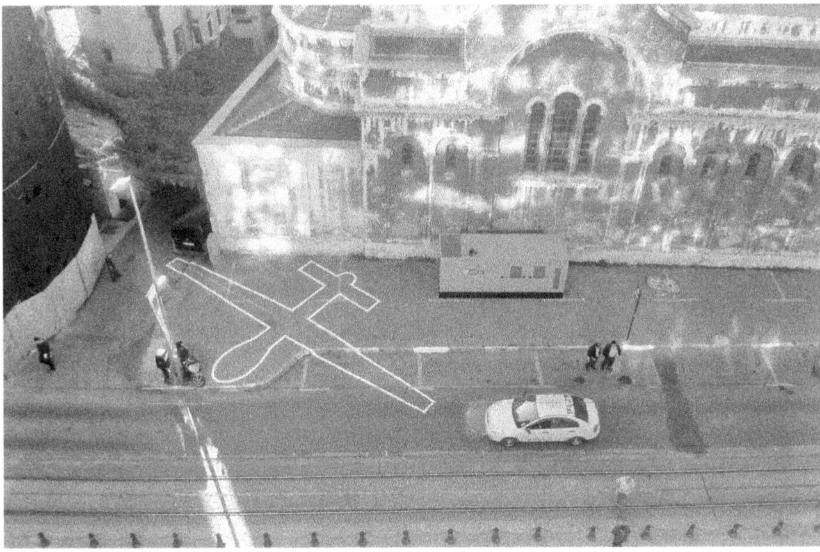

FIGURE 6. James Bridle, *Drone Shadow 002*, 2012. Commissioned for the Istanbul Design Biennale. Courtesy of the artist.

indicting the actors of a perverse network. However, for the viewer on the ground, these same marks struggle to coalesce into a meaningful gestalt and in fact appear destined to merge with the visual landscape of the city. This latter dynamic was driven home by a 2013 incarnation of the project for the Brighton Festival, which utilized a highway marking team to measure and paint the drone diagram on the seafront location of Madeira Drive. The familiarity of the orange cones, yellow safety vests, and road marking vehicles, which unceremoniously sprayed the road to indicate the silhouette of a Reaper drone, served only to magnify the capacity of this intervention to be buried within the everyday.

As Svea Bräunert describes: "In order to see the outline in its entirety—especially such larger drones as the Predator and Reaper—one must view it from above. Viewers thus are forced into a potentially compromised position as the view from above is also the view from the drone. And taking on its perspective can be both an uncanny and empowering experience."[17] The work thus turns upon a conflict between, on the one hand, the immediate experience of the drone outlines, and on the other, an uninhabitable mode of address that displaces experience to an elsewhere. On account of this bifurcated experience, the viewer's engagement with the drone in Bridle's piece is to a large extent contingent upon the photographic mediation of the scene, which not only preserves this otherwise ephemeral gesture, but more immediately grants the necessary distance from which to successfully read the image.

In its play between the visual omniscience of the aerial view and the fragmented testimonies of networked digital photography, the work reproduces the drone's interpenetration of hierarchical modes of power with diffuse and horizontal collectivities. This is exemplified by the kill chain itself, which as Derek Gregory describes "can be thought of as . . . a congeries of actors, objects, practices, discourses and affects, that entrains the people who are made part of it and constitutes them as particular kinds of subjects."[18] As each Predator drone requires around 185 personnel to operate, this expansive network does not simply remove the operator from the vehicle, but rather distributes its operation across a multitude of agencies. This disparate network includes "pilots" at Creech Air Force Base in Nevada, joint terminal attack controllers in Afghanistan, military lawyers at CENTCOM's Operations Center at Al Udeid Air Base in Qatar, and image specialists at the Distributed Common Ground System (DCGS) at Langley Air Force Base in Virginia, to say nothing of the host of inorganic actors that also populate the kill chain.[19] In light of this distributed mode of operation, the knee-jerk reaction to associate the all-seeing eye of the drone's aerial image with a consolidated and singular vantage point unravels, confirming Bruno Latour's claim that "the view from nowhere . . . also means that there is nowhere for those who hold it to realistically reside."[20]

Reiterating these relations, the drone in Bridle's work operates as a "partial object" in a dual sense. In the psychoanalytic tradition, Melanie Klein used the phrase to convey the processes by which the child forms relationships with discrete body parts of the other (usually the mother's breast) in anticipation of processing the whole.[21] In the context of databases, the same term is used to refer to those objects that have broken "invariants," conditions that typically remain consistent throughout the course of an operation. These artifacts have been partially reconstituted from the database but are not fully initialized by virtue of being separated from the rest of their data context. In both the psychoanalytic and computational usages of this term, the fragment is positioned toward a virtual future of full embodiment, one that emanates from an ecology of object relations and/or data attribution. The drone appears as a similarly incomplete object in Bridle's work, a shadow that is in expectation of an identity. As the activation of this image-object requires the image and its network, its coming into being produces a profane correlation between the corralling of data via the martial networks of drone violence and the algorithmic workings of a viral image shared across the distributed networks of social media.

The inspiration for Bridle's project was drawn from everyday digital media, specifically, the accidental shadows of drones that appear on Google Maps and other navigation platforms. Inadvertently captured by mapping satellites and/or commercial drones, these shadows are part of the optical unconscious of the everywhere war. As these representations render military technology both omnipresent and invisible, they reference a collective fetish logic by which the citizens of the Global North are able to disregard the reality of drone violence while remaining fully aware of its carnage. Collected on numerous websites and blogs, these shadows thus present a kind of Lacanian residue or stain within the otherwise seamless representation of the world, a reminder of the conditions that make this perfect legibility of space possible. However, as Bridle's work translates these shadows from the virtual space of the screen to the urban landscape, their impact is no longer purely symbolic. In this new environment, these images do not simply expose the presence of drones within the digital everyday, but rather invite a redistribution of the potentiality of drone networks by way of the digital camera. As the extensive online archive of photos suggests, *these* drone shadows desire to be returned to the screen by way of the photographic act. Presented as enigmatic urban pop-ups, they provide precisely the kind of photo op that fuels the image production of users of Facebook, Instagram, Flickr, Snapchat, and Instagram.[22] Through this deferral to the photographic frame and its privileging of circulation, the drone is hijacked by a new future orientation, a "counter-becoming."

In this regard, Bridle's drone shadows bear some similarities to Joseph Delappe's *Cowardly Drone* series (2013–), in which the artist downloads images of military drones from the Internet and adds the word *Cowardly* to the fuselage in the same color and font as the other official markings on its body (see figure 7). Most immediately, this defacement serves to call out the discrepancy between the safety drone operators, analysts, and commanders enjoy relative to their targets. In this regard, the work conjures anxieties concerning "killing from a distance" that go back at least as far as Homer, who bemoaned the use of archery in the *Iliad* on the very same grounds. However, more than the intervention itself, it is the virtual afterlife of this gesture that motivates

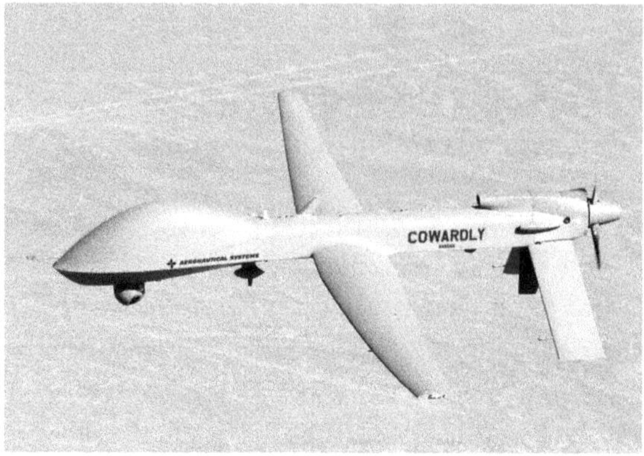

FIGURE 7. Joseph Delappe, *Predator Drone—Cowardly 2*, 2013. Modified digital image. Courtesy of the artist.

the artist's practice. Delappe explains that after the photos are altered, they are "uploaded to my website with basic titling information 'Predator Drone,' 'Reaper Drone,' 'Global Hawk Drone'—with the intention of having these images begin to appear in searches for information and images on drones occurs online. The works are intended as a subtle intervention into the media stream of US military power."[23] The introduction of these images into the online sphere serves not simply to circulate an oppositional message, but rather to provide the means by which future counternarratives might come into being. In this regard, the scope of the project extends beyond the unexpected appearance of Delappe's defaced drones in the results of a keyword search, though this interruption of the homogeneity of online worlds is not without significance. Rather, the work targets the underlying processes that shape the online experience itself, specifically, the actions of algorithms and content based image retrieval (CBIR) systems.

By way of their recursive operation, algorithms maintain a capacity to "learn" from user activity and modulate categories accordingly. In this regard, they do not simply infer but also partially produce the classifications by which their functions execute. At the same time, CBIR systems such as Google Image have begun to retrieve and organize images based on the actual content of photos rather than the metadata attributed to them by human users. With the visual information of the image made accessible to algorithms, the online dissemination of these photos allows the *Cowardly Drone* project to enter into the process of meaning-making at a base, operational level. In so doing, it does not so much expose or critique drone power, but rather call upon the internal dynamics of the network to produce "drones" as something other. In similar fashion, Bridle's empty forms are positioned toward a future activation, so much so that they refuse full presence in the now. However, by integrating the practice of photography into the work, the *Drone Shadows* series escalates the virtual dimension of this intervention, summoning both an archive of as-of-yet unmade images and the social networks that would partially determine these virtual presences. The question in the face of this work is no longer what alterations the artist makes to the image, but rather what kind of labor this prompt to photograph intimates. How does this shared temporality of making and viewing foster this infiltration of the database by the image?

Acknowledging the primacy of the network to the practice of digital photography, Kris Cohen describes the digital photographic operation in terms of a "minor intentionality" in which "photography produces photographs as effects, but isn't superseded by them and doesn't even seem to be the necessary cause of them."[24] This weakened causal relationship between photographic practice and the images it produces provides the conditions of possibility for the collapse of shooting and sharing images. While maintaining a fragile spatiotemporal separation, these activities come to not only largely determine one another, but also take place within an expanded continuum of practice. This entanglement forms the basis for what Cohen calls a "new categorical richness" of the photographic moment under digital photography. Here the "frozen instant" gives way to an incessant, ecstatic doing as images are recontextualized, redefined, and reconstituted within the networks of social media. Summarizing these sentiments, Caterina Fake, the founder of Flickr, explains: "The nature of photography now is it's in motion. It doesn't stop time anymore."[25] In similar terms, while the drone in Bridle's piece only becomes legible via the image, the act of photographing does not constitute an end in itself. Confirming the inability of locating "the work" in its totality, this practice conjures a futural dimension in which the photographic gesture exists as an ongoing iterative process. One might describe this unique status of the work and its images in terms of what Erin Manning has called "preacceleration," the momentum of a movement that has not yet taken place but manifests as the instability of form.[26]

Drone Shadow's eschewal of complete or definitive representation compromises the definitive presence of the object in order to expose it to the potentiality of digital network. In this distributed ecology, the still image breaks down and a disruptive futurity is summoned. This strategy echoes what Artie Vierkant and others have referred to as "post-Internet art," a mode of production that focuses on the distributed nature of authorship, the porous boundaries of the museum or gallery, and the infinite malleability of image-objects. As Vierkant describes, this work is built upon the acknowledgment that

> nothing is in a fixed state: i.e., everything is anything else, whether because any object is capable of becoming another type of object or because an object already exists in flux between multiple instantiations. . . . In the Post-Internet climate, it is assumed that the work of art lies equally in the version of the object one would encounter at a gallery or museum, the images and other representations disseminated through the Internet and print publications, bootleg images of the object or its representations, and variations on any of these as edited and recontextualized by any other author.[27]

To return to the founding question of this chapter—namely, how one can photograph the drone and its corresponding relations of power in light of its distributed status—Bridle's *Drone Shadows* poses an interesting proposition. In utilizing the potentiality of "the networked lens" to reproduce the distributed status of the kill chain, the work effectively reconfigures the historical operation of the camera so that its images take on what Stefan Heidenreich describes as the "liquid surface" of post-Internet art.[28] It does so by way of its skyward orientation, which strategically reiterates an enduring logic of wartime aerial photograph in its insistence that "close-up images [can] not be understood unless they [are] placed in a context with a much wider or distant view."[29] The lack this gesture produces for those on the ground summons the digital camera to grant content to these empty shadows and thereby invite the ontological movement of the network. As a result, the singular image gives way to a shifting multiplicity, just as the individual creator gives way to a dispersed multitude. These relations are exacerbated by Bridle's latest addendum to the project, which features a how-to-manual that allows anyone to reproduce drone shadows in their city. With installations appearing in Detroit, São Paulo, and other locations around the world, this gesture has initiated a global dispersion of drone shadows. The ongoing *détournement* that has followed suggests that the same super-pliable networks that grant drone power its ability to reconfigure and react in real time also open these systems up to infinite transgression. In this way, the work dramatizes not only the structural overlap between the everyday mediasphere and the martial networks of the everywhere war, but also the vulnerability of drone power to what Latour calls the "reversibility of [its] foldings."[30]

While Bridle's work prompts the camera to peer down from above, the projects considered in the next subsection gaze upward into the atmosphere in order to disclose the inner workings of drone power. In conjuring familiar tropes of the ineffable, these photographs establish a correlation between the unknowability of the natural world and the limitlessness of drone power. However, rather than being a simple metaphor, this relationship is one of strategic interpenetration, whereby forces of nature act as collaborative agencies in the implementation of everywhere war. By enfolding the natural world into the algorithms of martial networks, these works by Trevor Paglen and Scott Patrick Wiener acknowledge the sky as medium, a means of producing representation that echoes the operation of the camera itself.

Nature as Vertical Empire: Trevor Paglen and Scott Patrick Wiener's Untitled Series

As the seat of a war without geographic boundaries, the sky serves as a critical medium of the global counterinsurgency. Its integration into networked warfare proceeds not from the historical drives of claiming territory or creating footholds within airspace, but rather from the creation of a depersonalized and parenthetical space through which law and national identities can be strategically suspended. Drone warfare realizes this ellipsis by shifting the mode of operation from an attempt at mastering the skies from without via strategic relations of visibility or overt displays of power to a co-opting of the atmosphere itself. In this regard, drones exemplify a larger operation, which Hill refers to as "aerial empire," a convergence of technological and environmental forces that embeds power relations within the natural world:

> The term aerial [in this context] should signal not only the destructive potential of the atmosphere, catastrophic climate change, and by extension, the coming resource wars, but should also pertain to the spatio-informational armament, endless ensembles of antennae, satellitic vision, those super-sites of knowledge transmission, data analysis, and tele-connectivity that re-orient conventional military arts. . . . In this "new" ecology of war . . . the epistemo-military arts are developing techniques that incorporate a symbiosis of agencies—biological, atmospheric, geological, and mechanical—that from any other perspective than war would be revealed as the suicidal miscellany of a planet abandoned to siege.[31]

This expanded category of war posits a startling symmetry. Just as carbon emissions reallocate resources and reaffirm pervasive hierarchies through "natural" disasters, so does networked warfare exercise geopolitical power through an uncanny atmospheric agency of its own. In both scenarios, an assimilation of the outside unifies opposing drives so that hierarchical modes of power are ultimately reaffirmed via an enfolded other, even while the two phenomena appear distinct and unrelated.

In the case of drone warfare, the engine for this ontological enfolding is a tactical program called the kill box. Originating in the 1990s with the joint operations doctrine and revamped in 2009 as the air-tasking order (ATO), the kill box is articulated by an extensive daily document that outlines the next twenty-four hours of operations for each theater of operations.[32] As Derek Gregory describes, this program divides aerial space into a series of nested cubes that operate as scalable fields: a "cell" is subdivided by four "quadrants," which are in turn divided into five "keypads." Permission to fire is granted within a given number of cells and restricted by a specific window of time.[33] On this basis, certain zones are labeled as no-fire areas (NFAs), restricted operations areas (ROAs), airspace coordination areas (ACAs), and so on. However, as designated cells open and close throughout the mission according to "target cycles," these spatial and temporal designations remain fluid.[34] In this way, the kill box forms a dynamic container of possible actions that effectively correlates dispositions, identities, and behaviors to atmospheric parameters of space and time.

Despite relying upon twenty-first-century computational processes and digital networks, these relations of the kill box draw from a long history of the sky as wartime medium. John Durham Peters uses the term "sky media" to refer not only to the formative role that clouds have played in pushing representation beyond the symbolic in the history of art, but also to the meta-operations by which these bodies determine relations of visibility. In overseeing both the visual access between land and sky and the quality and quantity of light that reaches the eye, the "natural" processes of the sky for Peters represent a primary form of image making.[35] With the advent of flight, these same relations made the air a critical collaborator in the operations of war. As Peters explains, "since the utility of spy planes depended on clear weather, espionage was networked with meteorology, another form of reconnaissance. Clouds cloaked targets and learning to read clouds was a spin-off of the urgent need for reassessing nuclear arms stockpiles."[36] To this end, the

US Army developed the Electronic Numerical Integrator and Computer (ENIAC) after World War II to not only "calculate[e] bomb trajectories and kill ratios but also [provide] the computational power necessary to produce [. . .] mathematically based, multi-layered atmospheric forecasting."[37] However, for this system to be effective it would require a stream of real-time data that would exceed the maximum altitude (and fuel supply) of conventional airplanes. When 16mm footage captured by an Aerobee Rocket revealed a tropical storm system that had eluded forecasters on the ground ten miles below, the military realized that a permanent observation platform within the exosphere was the answer. In response, the Research and Development Corporation (RAND) was requisitioned to create what would become the first operational weather satellite system. The Cold War meteorological network that followed rendered the sky not simply knowable, but predictable. In so doing, it initiated a paradigm shift in which passive observation of atmospheric conditions gave way to active intervention. The most notorious example of the latter is the environmental modification operations (ENMOD) used by the US military during the Vietnam War. Operation Popeye, as it was called, utilized C-130 and F-4C planes to seed clouds with lead iodide and silver iodide in order to extend the monsoon season and thereby wash out the Ho Chi Minh Trail and other thoroughfares.[38] Just as important as the acidic rain that fell was the cover these synthetic storm clouds provided for the South Vietnamese commando units as they moved into North Vietnam.[39]

While operations of "weather modification" understood the natural world as a medium that could be manipulated in order to maximize the efficacy of military operations, a simultaneous trajectory of martial tactics sought to weaponize the air itself. This collapse of martial and natural processes forms the basis of what Peter Sloterdijk calls "atmo-terrorism," a mode of twentieth-century warfare that did not so much target the body of the enemy as the environment upon which it relied. Born out of the stalemate of trench warfare in World War I, this strategy, which evolved from pesticides to chlorine gas to gas chambers to the radioactivity of the atomic bomb, embeds its assault within nature, co-opting the symbiotic relationship of the body to its environment, especially the air.[40] While environmental warfare and weather modification were eventually banned by international treaty, the pursuit of a collaborative relation with natural forces would continue to structure the wartime relation to the sky. In fact, it is precisely this historical correlation between the sky, visibility, and martial violence that resurfaced in October 2013 when a thirteen-year-old boy from Pakistan named Zubair testified to Congress regarding the US drone attack that sent shrapnel through his legs. He explained: "My grandmother and I used to share a love of bright blue skies. We have many of them in Tappi, the village where I live. . . . I no longer love blue skies. In fact, I now prefer grey skies. The drones do not fly when the skies are grey. . . . When the skies brighten, the drones return."[41] Most immediately, this articulation of drone subjectivity as fear of the sky attests to an ongoing entanglement of martial agency with the atmosphere that has accompanied military conflict for the last century. However, the personification of the sky one reads in these remarks also acknowledges a specific constellation that takes shape under drone power, one that expands martial violence beyond the singular event into a generalized condition of the air. Drone skies achieve this status by augmenting historical modes of environmental intervention with an operational program that enfolds the attack into the atmosphere at a symbolic level. Through this union of material and immaterial forces, the attack comes to take place within an expanded temporal continuum that conjoins the immediacy of the strike with the "slow violence" of environmental toxicity.[42]

Lisa Parks argues that the drone's mode of "vertical mediation" exceeds the confines of the screen and in fact encompasses a "vast expanse that extends from the Earth's surface, including the geological layers below and built environments on it, through the domains of the spectrum and the air to the outer limits of orbit."[43] For those who live under drones, this connection between the UAV and elemental phenomena such as "ground surface, air, spectrum, orbit, labor and energy" is perhaps most legible in the environmental legacy of the strike itself.[44] Numerous studies conducted on the environmental effects of the US invasion of Iraq have shown that the same missiles used by drones have left depleted uranium in the groundwater, which remains contaminated decades after the conflict. In Waziristan, the epicenter of the US drone war, a number of residents have developed skin disorders, respiratory diseases, and blood cancer since the strikes have begun. Increasingly, experts are drawing connections between the health issues of the region and the environmental devastation of Fallujah, where cancer, leukemia, and birth defects have skyrocketed in the years following US military actions. In Gaza, doctors have observed that subjects who live near drone strikes tend to exhibit excessive bleeding after surgery, despite ample administration of plasma.[45] Perhaps the most damning evidence of the toxic atmospheric footprint of drones is supplied by the World Wildlife Fund, which reports that the presence of migratory birds over

Afghanistan has dropped by 85 percent since the regular deployment of these missiles in the area.[46]

Far from being unintended side effects, the environmental consequences of strikes mark what Hill describes as the "co-operational" relationship of drone power with nature, the tendency of the former to regard the latter as a "functional part of the atmospherically embedded war machine."[47] What differentiates drone power from previous instances of atmo-terrorism is the way in which the underlying logic of these interventions is reproduced within the symbolic systems of the screen. As the kill box eschews individual targets in favor of conditions and atmospheric parameters, the sky no longer functions as an empty container in which violence takes place. Rather, it acts as a distributed agency that materializes martial force within the diagrammatic spaces of the operational sphere as well as the lived experience of the atmosphere. In this way, the symmetry produced by the drone's unique version of atmo-terrorism articulates a dual temporal axis by which the immediate violence of the Hellfire missile reverberates across an elongated narrative of toxicity. In the face of this interpenetration of the violence of war with the sky, is it so naïve to assume that the attack emanates from the sky, to read the clear sky as communicating an impending strike? This would seem to be precisely the proposition of Paglen's *Untitled* and Wiener's *Untitled 2 (Spies in the Sky)* (2013) series, which employ a series of art historical and popular culture references to disclose this collaboration between drone power and the atmosphere.

Paglen's photographs feature large, indistinct swaths of horizonless skies (see figure 8). In this space, shadows disappear and light loses directionality, leaving a perfectly uniform and flat illumination that suggests the light-emitting diode of a screen rather than the fickle rays of the sun. As these monumental abstractions of clouds, light, and color fill the frame with an almost impossible vividness, the viewer has the sense of an intrusion when the tiny UAV comes into view. At first glance, its presence appears coincidental, as if the photographer unknowingly captured these aerial machines while innocently searching for the perfect arrangement of clouds. However, as this same interruption is repeated throughout the series, the viewer eventually begins to anticipate this presence and actively searches these large expanses of sky (the images are forty-eight by sixty inches) for the drone. In the course of doing so, it becomes clear that like the satellite images of UAV shadows that fall upon unsuspecting roads and open expanses of land in Google

FIGURE 8. Trevor Paglen, *Untitled (Reaper Drone)*, 2013. C-print. 48 × 60 inches (121.9 × 152.4 cm). Courtesy of the artist and Metro Pictures, New York.

Maps, the drone in these works is a stand-in for a larger system of relations, which both exceeds and structures the photograph in which it circulates. It is by way of these relations that the expansive skies that surround the drone in these images become entangled with martial networks.

Through their grand scale and overall sense of unbridled nature, the photographs of Paglen's *Untitled* series forges a rather tongue-in-cheek deployment of the sublime landscapes of Romantic painters such as J.W.M. Turner and Caspar David Friedrich. In these latter works, the sky exemplified the Kantian sublime in its affirmation of the inadequacy of human reference in the face of the absolute of nature. However, in Paglen's images the same extravagance is rendered oddly legible, coherent, and present. Despite having relinquished the horizon and any explicit signs of human presence, the sublime's conventions of illegibility seem to take on the certainty of tactical space. In recouping signifiers of unknowability as the field markers of martial networks, these photographs correlate drone power with an assimilation of the outside, defined here as that which exceeds both symbolic systems and human agency. With the sky comprising a medium that both precedes and makes possible communication, the denotive aspect of these photographs is strategically undercut by what Sean Cubitt describes as a flow "which precedes all separations, all distinctions, all thingliness, objects, and objectivity . . . [that which] precedes the separation of the human from the environmental."[48] In this way, the anti-ontology of Paglen's clouds not only renders the virtual force of the signature strike legible, but also presents the air as partial origin of its violence. As this dialogue of atmospheric enfolding is extended to the ground below in Wiener's *Untitled 2 (Spies in the Sky)* (2013), it is the underlying conventions of the genre of landscape that serve as the grammar by which the earth is shot through with the logic of the kill box.

Wiener's monumental, black-and-white images are screen captures from a Time/Life documentary on modern drones entitled *Spies in the Sky* (released on VHS in 1999). After being extracted from the video, these stills were output as small-scale inkjet prints, then photographed with high-speed film and enlarged to forty by sixty inches. The resulting images are grainy, picturesque, and haunting (see figure 9), as if the nineteenth-century photographer Julia Margaret Cameron had photographed the landscape from a precarious hot air balloon. Wiener describes the motivation behind the selection of these images in terms of two criteria. First, they had to be harvested from those portions of the show that "acted as placeholders . . . and were meant only to illustrate the specter of surveillance from above."[49] In other words, these stills were to be taken from the "B roll," the footage that is shot offsite or procured from the archive and intercut with the action in order to add ambience or depth to the diegetic present. Second, the images had to "bear a formal relationship to historical landscape aesthetics and bring to mind the depiction of the sublime in painting (German Romanticism, Dutch landscape, Hudson River, etc.) and photography (O'Sullivan, Ansel Adams, New Topographics, etc.)."[50] Together these criteria ensure a layered mode of appropriation in which the artist not only extracts stills from a film in which they already exist as a kind of found archival object, but does so in order to replicate an enduring genre from the history of art. As a result of this layered web of chronologies, references shift and move with disruptive uncertainty within the image. In the oddly familiar landscapes below, one sees traces not only of Turner and Friedrich, but also of the Hudson School. In fact, one image from the drone's camera seems almost to reproduce verbatim Thomas Cole's seminal

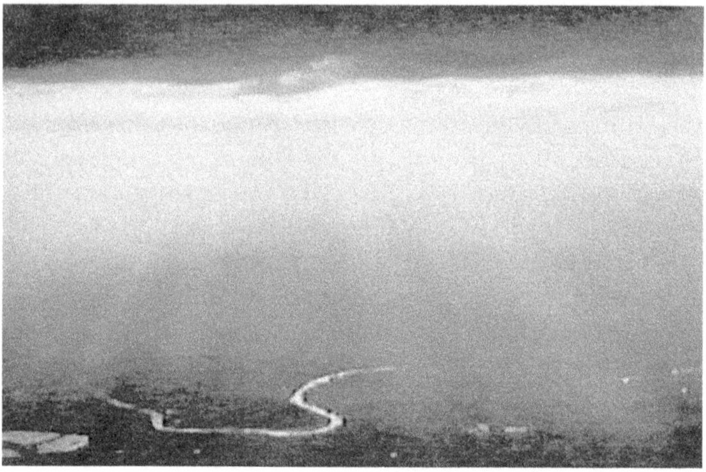

FIGURE 9. Scott Patrick Wiener, *Untitled (Spies in the Sky)*, 2013. Archival inkjet print. 40 × 60 inches (101.6 × 152.4 cm). Courtesy of the artist.

work *The Ox Bow*. As the viewer is prompted to see from the drone's-eye view, these references become interpenetrated with the instrumental logic of the kill box. The Poussin landscape—that instantly recognizable set of conventions in which meandering lines of ancient rivers and worn pathways lead the eye seamlessly through the pictorial planes of foreground, midground, and background—here becomes a process of stacking the horizon with geometric volumes through which potential strikes can be actualized.[51] The boundaries of streams and lakes partition off sections of the landscape, circumscribing a series of cartographic units through which shifting boundaries momentarily stabilize as target. As these conventions provide the architecture for the scalar logics of the kill chain, they simultaneously dramatize the breakdown of the spatial organization and overall singularity of these represented worlds, pushing the image toward a relational character in which the futural dimension of the kill chain manifests as unformed potentiality.

These relations are made possible by not only the act of appropriation, but also the unique aesthetic of the screen capture. The screen capture can be distinguished from the film still based on the relationship it maintains to its source material. The latter enacts what might be called a soft betrayal of its source, which by virtue of being a "moving picture" cannot be adequately conveyed by this disconnected photographic artifact. As a result, the still, especially its promotional variety, demarcates an impassable outside, a lure which in attempting to encourage a future experience of the film explicitly denies such entry in the present. It is for this reason that this form proves to be such an effective marketing device, as it effectively requires the purchasing of the ticket or, more broadly, the input of the user in order to divulge its content. In this way, the still demarcates a lack that only the act of consumption can bring fullness to. The screen capture, on the other hand, operates via the logic of the thumbnail, which posits its images as a portal more than a partial representation. In offering access to missing content, the images are necessarily entangled with the content they reference. The whole and the fragment move through one another in the thumbnail and are in fact made mutually constitutive of one another by the sharing economy. Through Wiener's layered mode of appropriation, the nesting aspect of this relationship comes to echo the scalar logic of the kill chain, allowing the ebb and flow of the drone's powers of world-making to be made visible in the image.

Though not relying upon screen captures, Paglen's *Untitled* series nonetheless manifests a similar logic between images. Here the repetition of stock aerial scenes of identical size and shape prompt an infinite interchangeability, producing a tenuous unity that is partial and dynamic. Individual photographs of the series align themselves with near-perfect replicas as if responding to an image search, invisible fields of metadata uniting otherwise disparate spaces and times in a single thickened present. As this collection of placeless images bears all the signs of a logic that accumulates rather than narrates, compiles rather than documents, harvests rather than represents, the *Untitled* series collapses the network's powers of deformation with the sublime force of nature. Through these relations, the atmosphere materializes as a cloud of virtuality, a screen for the infinite variability of the martial network, which undercuts the image's claim to definitive or singular representation.

Achille Mbembe argues that colonial power rests not only upon the division of the land according to natural features (hilltops, water sources, etc.), but also on a correlative "separation of airspace from the ground."[52] This latter process is set in motion by an exhaustive cataloging of the sky, which overcodes and reconfigures the earth below so as to make possible the hierarchical relations that undergird occupation: "Even the boundaries in airspace are divided between lower and upper layers. Everywhere, the symbolics of the *top* (who is on top) is reiterated. Occupation of the skies therefore acquires a critical importance, since most of the policing is done from the air. Various other technologies are mobilized to this effect: sensors aboard unmanned air vehicles (UAVs), aerial reconnaissance jets, early warning Hawkeye planes, assault helicopters, an Earth-observation satellite, techniques of 'hologrammatization.'"[53] While the kill box disconnects military action from the potential resistance of the ground below (national boundaries, legal restraints, and discrete identities of actors and victims of drone violence), it also enfolds violent agencies within the airspace via the logic of atmo-terrorism. In this way, the cleavage between land and sky that has historically marked the operations of colonialism and its relations of "vertical sovereignty" is rendered not only partial but reversible. The sky is no longer a passive medium to be acted upon, but rather an agency to be worked through and in concert with. Through its deployment of the trope of the unrepresentable, unique modes of appropriation, and interchangeability of images, the work of Paglen and Wiener performs this assimilation of nature into the drone's ecology of violence. In so doing, the still image diagrams a mode of power in which strategic instances of recentralization corral the immanent relations of the network as natural force.[54]

CONCLUSION: PHOTOGRAPHIC BOX, DATABASE SUN

The kill chain is comprised of a series of nested boxes, an army of matryoshka dolls that collapse into one another in order to actualize an event and unfold into a multiplicity to maximize responsivity. As this field of molar entities oscillates between micro and macro instantiations of a given program, it produces a kind of Droste effect in which objects, agents, and events inside the frame are simultaneously outside and vice versa. Ultimately, this collective enfolding manifests the force from without that underlies broader relations of tactical animism discussed in the previous chapter. However, in order for this force to take hold, each individual component of drone networks must also internalize the global logic of drone power. For this reason, the spaces, processes, and agencies of the kill chain represent something more than the composite parts of drone networks. They are in fact distillations of the larger program of everywhere war. As the discussion of drone cinema in chapter 5 illustrates, the cockpit or "box" of drone operators exemplifies this condition. Marking a copresence of near and far, real and virtual, artificial and natural, its modular and nonsingular status articulates a space whose boundaries redistribute as much as they contain. The impossible simultaneities of drone warfare run through these Conex boxes even as their darkened interiors appear to partition off life outside. However, the most glaring and immediate confirmation of these relations is the drone itself. As both discrete object and collective agency, material presence and virtual elsewhere, its conflicted mode of presence serves as a metonymic stand-in for the nesting logic of martial networks. It is for this reason that the drone proves to be such a difficult subject for a medium that has historically privileged stillness and the authoritative status of singular images.

In presenting the drone in terms of a prompt to photograph, Bridle's *Drone Shadows* installations seek to harness the disruptive movement of the image as it circulates through distributed networks. Through a layered web of appropriation, the work of Paglen and Wiener, on the other hand, reproduces the relations of the kill box within the frame. While each body of work offers distinct responses to the tensions that photographing the drone present, a shared strategy nonetheless comes into view when read in relation to the nesting logic of drone power. Collectively, these projects suggest that photographing the drone entails presenting the image as one box within the sequence of kill boxes that animate drone violence, i.e., conceiving of its frame as porous and its contents as subjected to flows that undercut ontological stasis. This program of the "photographic box" is clarified by two works of drone art: Mahwish Chisty's *Kill Box* (2016) and Scott Patrick Wiener's *Visibility Tests for Sky* (2011–2015). While the former succinctly conveys the strategy behind the works discussed in this chapter, the latter drives home the consequences of such an intervention for the still image.

Upon returning to her native Pakistan for a visit in 2011, artist Mahwish Chisty was struck by a heightened US military presence in the daily life of friends and family members. In response, she began to interrogate visual representations of the counterinsurgency, focusing in particular on the drone. At the time she found not only a relative absence of images of the latter in the digital sphere, but also a lack of faithfulness in the few photographs that did exist. She explains: "I started to question the photographs that we could see and find online. I started to think about who's the photographers, where is the camera placement, where is the photograph from and hence questioning the authenticity of the photograph itself."[55] While researching these issues, Chisty learned that the 2009 representation of the RQ-170 drone featured on Wikipedia was itself a product of a similar conundrum. Using the only four photos of the drone that were available online at the time as his guide, creator

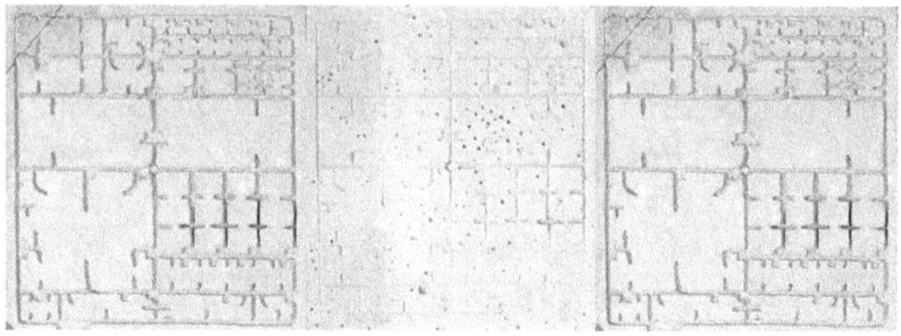

FIGURE 10. Mahwish Chisty, *Kill Box*, 2016. Clay tablets. Courtesy of the artist.

Mark Miller had produced a generalized representation that soon become an iconic image. In the process, he initiated what Chisty came to regard as a recurring play between presence and absence in the representation of drones, a dynamic that she directly connected to the discrepancy between the prevalence of drones in her grandmother's life in Pakistan and the scarcity of images of these weaponized vehicles online. This narrative came to serve as the basis for numerous paintings, videos, and models of Predator drones, which incorporate local motifs such as Pakistani truck art and the war rugs of Afghanistan in order to reclaim these missing images of drone violence.

Within this body of work, it is Chisty's *Kill Box* that is most relevant to the current discussion.[56] The piece is comprised of a clay tablet with what appears to be an ancient script pressed into its surface (see figure 10). Despite conjuring cuneiform-laden artifacts from the region's rich historical past, the object's enigmatic design is actually an impression of the plastic frames that hold the parts of a toy drone. This conflict of histories is magnified by the fact that the rectilinear script this mold produces seems also to mimic the diagrammatic data that appear on the screens of the kill chain (as the artist explains, the impressions are "amazingly close to some of the images that I've seen of drone operators on their monitors").[57] From this web of contradictory associations, a discursive and technical constellation emerges that not only correlates this disruptive movement with the medium of photography, but also reveals the larger program of drone art in the process.

At its most basic level, *Kill Box* attempts to give presence to an absent object by way of an inverse image. In so doing, the work reiterates what was the basic logic of photography for more than a century and a half.[58] In fact, the first negative-based photographic system was developed by W.H. Talbot in part to document the clay tablets in the British museum.[59] In attempting to "cause . . . natural images to imprint themselves durably, and remain fixed upon the paper," the calotype essentially reimagined the inverse impressions of these ancient media as proto-photographic gestures. Fellow photographic pioneer Nicéphore Niépce acknowledged this relationship by regularly referring to the photographic plate as a tablet in his writings. Soon afterward, it became widely accepted to describe the medium in precisely these terms, i.e., as "the silver tablet," "the blank tablet," and so on.[60] Given both the entangled history that unites these media and the fact that the artist's work on drones was born out of an encounter with the still image, one cannot help but regard this "ancient" tablet in terms of the photograph.

And yet, rather than a photorealistic representation of the drone, the work presents only vague impressions of its component parts. The tablet in fact ensures that these traces fail to summon a distinct and singular object, not only by bringing together conflicting historical and technical references, but also by flipping the photographic logic of the negative impression. In presenting the inverse inscription as the final image, *Kill Box* sabotages the interplay of presence and absence that typically undergirds the photographic image, effectively summoning the architecture of the drone without the drone. Most immediately, the work utilizes these dynamics to convey the lack of presence and overall sense of inauthenticity that the artist first observed in photographs of the drone. However, when read in conjunction with the work's evolution and web of historical associations, these same features seem also to function as a call for the photographic medium to invert its conventional operations so as to successfully engage with the drone. The transformations that such a program requires of the still image are dramatized by Scott Patrick Wiener's *Visibility Tests for Sky*.

Wiener's work presents a montage of repeating images of a drone surrounded by varying shades of blue, which are identified as "electric blue," "sky blue," "azure blue," "snow white" and so on (see figure 11). One is tempted to read this progression in chronological terms, as each subtle shift in tonality seems to suggest the temporal interval of a comprehensive cycle, which moves through absolute darkness, full sun, and back again. And yet the progression formed by this gradient of color begins and ends abruptly, without closing the circle. Nonetheless, an order does seem to persist. Rather than the sun, it is the drone which, as Wiener puts it, "progressively disappear[s] and reappear[s]" across the series.

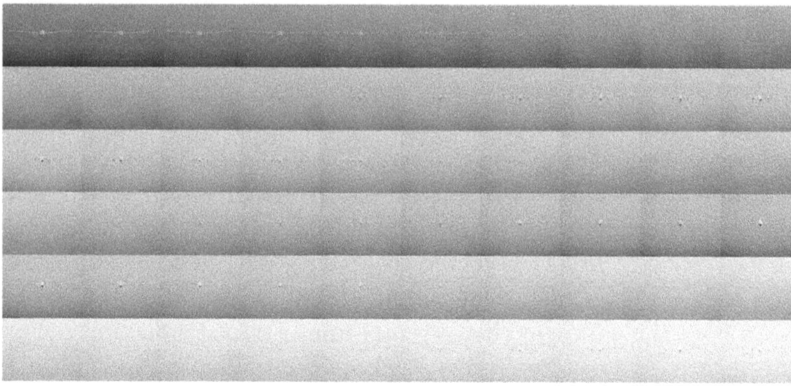

FIGURE 11. Scott Patrick Wiener, *Visibility Tests for Sky*, 2011–2015. Courtesy of the artist.

While the drone appears in the same position in each image, the conditions that surround it shift between iterations. The result is that the movement of this vehicle through space is supplanted by an internalized movement between states. In dramatizing this variability of conditions, identities, and visibility, this sun no longer charts a continuous path across the sky, but instead oversees a variable movement or modulation through a given set of parameters ("electric blue," "sky blue," "azure blue," "snow white"). This sun correlates its movements to GPS coordinates, emits light according to hexadecimal codes, and becomes legible according to the oscillation of conditions rather than the articulation of linear time or the production of navigable space (there is no origin of this light, no shadows that give direction). This is the light of the database sun articulated by Umbrico's installation, discussed at the beginning of this chapter, a light whose life-giving properties extend beyond the biological and into the formative processes by which identities, events, and places come into being. In the face of this light, the solidity of the photograph breaks down, becoming the site of infinite variability rather than of static representation.

All of this is contingent upon the symbolic labor of the viewer, who is effectively called upon to produce this movement. After all, the gradient formed by these images inevitably reads as a paint swatch used in renovation or, even more relevant, the "color picker" of the popular Adobe image-editing program, Photoshop. In the face of this familiar interface, the image self-presents as medium rather than representation. Like the screen of the drone operator or the coordinates that determine engagement, this photographic box does not simply accommodate variable content, but rather renders content variable. As a result of this status of the image, the totality of the larger montage becomes infinitely reversible, open to acceleration or deceleration. In this way, the work discloses the fundamental assumption of drone art, namely, that the same limitless variation that circumscribes drone violence also opens it to infinite revision.

CHAPTER 3

From the Ground Below

Spotting Industries, Smartphones, and the Post-Panopticism of Drones

Fieldwork conducted in Pakistan by the *Living under Drones* project confirms that drones maintain an overt and at times even conspicuous presence in the lives of their subjects.[1] A young father told interviewers that the drones "make it difficult to sleep. They are like a mosquito. Even when you don't see them, you can hear them, you know they are there."[2] Another respondent explained: "We can't go to the markets. We can't drive cars. When they're hovering over us, we're all scared. One thinks they'll drop it on our house, and another thinks it'll be on our house, so we run out of our houses."[3] These reports of the everyday presence of drones are echoed by David Rohde, a former *New York Times* journalist who was held hostage by the Taliban for seven months in Waziristan. He recalls that throughout his captivity, "American drones were a frequent presence in the skies above North and South Waziristan. Unmanned, propeller-driven aircraft, they sounded like a small plane—a Piper Cub or Cessna-circling overhead. Dark specks in a blue sky, they could be spotted and tracked with the naked eye."[4] These firsthand accounts confirm that drone power is not only capable of operating outside of imaginary or "internalized" modes of nonverifiable presence, but in fact regularly traffics in material, embodied, and phenomenological presence.[5] Why is it then that academic and popular discourse consistently associates drone power with relations of voyeurism and invisibility?

In a 2013 issue of *American Quarterly* dedicated to visual culture and the war on terror, Matt Delmont summarizes this familiar refrain when he attributes the "deadly power" of the drone to its "twin claims to visual superiority: the ability to see and to resist being seen."[6] Chamayou similarly attributes the ethical lapses of the drone operation to the ability to "see [. . .] without being seen," a scenario he regards as confirming the conclusion of the Milgram experiment that it is "easier to harm a person when he is unable to observe our actions than when he can see what we are doing."[7] A similar correlation of drone power with the unseen forms the basis for the title of a 2016 exhibit held at the Kemper Art Museum, *To See without Being Seen: Contemporary Art and Drone Warfare*. While invisibility comprises a critical component of the UAV's tactics and operations, the ascendancy of this panoptic discourse in theorizing drone power carries with it a host of dangers. Not only does this conceptual framing tend to obscure the nonvisual elements that contribute to the drone's "mode of address," but it also elides the gestures of revelation that prove equally critical to its operation. By negating the direct forms of presence that drones articulate for subjects on the ground, such readings risk universalizing the position of safety of those untouched by its violence. As a result, drone power comes to occupy an abstract elsewhere, which both conceals and enables its violence. The correlative understanding of instances of drone presence as momentary glitches or accidents within this regime of invisibility proves equally suspect. Considering that the Predator is capable of flying at fifty thousand feet, outfitted with laser-guided, long-range munition systems and high-resolution cameras that can read a license plate from two miles up, these gestures of self-revelation obviously cannot be explained in purely operational terms, nor can they be written off as simple mistakes.

In order to fully acknowledge the experience of those on the ground, familiar tropes of panopticism and invisibility must be read alongside the drone's active pursuit of presence. From this perspective, it becomes clear that conventional dynamics of surveillance do not simply coexist alongside an opposing logic of explicit presence, but that drone power in fact relies upon a strategic oscillation between panoptic modes of invisibility and overt revelation. As these repeated performances of presence and absence correlate to the kill chain's internal movement between the actual and possible worlds, these gestures act as the performative equivalent of the relations of tactical animism, which grant the kill chain the ability to interpenetrate manifest reality with the not-yet-present of virtual force. By virtue of this correspondence between the presence of the UAV and the futural vector in which the strike appears as inevitability, the tactical operations of the drone become deeply entangled with the psychic processes of those who live under its shadow. Indeed, these relations play a critical role in the military's pursuit of "full spectrum dominance," a strategy that understands the psychological dimensions of combat as inseparable from aerial, nautical, cyber, terrestrial, and extraterrestrial fronts.

In "Inhibitions, Symptoms and Anxiety," Freud distinguishes "signal anxiety" from its less-evolved counterpart, automatic anxiety, on the basis of its anticipatory nature. It is not the loss of the object that produces the former condition, but rather the fear of the impending trauma this loss provokes. As the internal movement of drone networks increasingly manifests as the apprehension of a nonpresence that might always assert itself in actual terms, the Pakistani psychiatric community has begun to see the rise of a drone-related condition aptly referred to as "anticipatory anxiety." Following Freud, one might break down this anxiety in the following terms: a "signal" (the drone) announces a "danger-situation" (its implicit strike), in response to which the subject "anticipate[s] the trauma and behave[s] as though it had already come" ("anticipatory anxiety").[8] In this case, however, the "as if" does not require the fantastic mechanism of neurosis (nor the governing power of the ego) to attain the status of reality, but rather draws its currency from the inner working of drone power itself. After all, this already present nature of possible trauma, which for Freud defines the anxious state, is precisely the logic of the signature strike, which utilizes the martial database and its algorithms to organize worlds around violence rather than the reverse. In this regard, the oscillation of the drone between presence and absence does not simply conjure the possibility of violence, but rather reaffirms the presence of an impending world in which such an attack appears inevitable. For those on the ground, its unique mode of address offers confirmation of anxiety's illusion, testifying to the ease with which the "as if" can pass over into the substrate of the real.

Drone art engages with these relations by interjecting the visual and sonic experience of the UAV from the ground. In so doing, it unsettles both the discourse of invisibility and the asymmetric power relations that it subtends. Exemplary of this dynamic is James Bridle's *UAV Identification Kit* (2012) and Gonçalo F. Cardoso and Ruben Pater's *A Study in 21st Century Drone Acoustics* (2015), which in unique ways attempt to destabilize the directionality of drone power by mobilizing wartime practices of "spotting." If drone targeting momentarily stratifies immense waves of data in order to materialize sovereign power as force, then these works destratify this gaze in order to reflect back a hyper-fluid and limitless variation. Bridle's *Dronestagram* and Josh Begley's Metadata+ app continue this strategy in the context of the mobile device. Drawing out the interpenetration of civilian and military networks that comprise communication satellites, these interventions seek to interject drone violence into the everyday and thereby destabilize the geographic distinctions upon which its asymmetries rely. Collectively, these works utilize 3-D printing, crowdsourcing, and locative media to amplify the contingent status drones and their targets maintain within martial networks. In the process, the illusion of surgical precision crumbles, and the collapse of seeing and striking that undergirds drone vision becomes synonymous with indiscriminate violence.

THE IDENTIFICATION KIT AS OBJECTILE FORM

In his *Identification Kit*, Bridle uses 3-D printing to produce miniature models of three contemporary military drones—the MQ-1 Predator, the RQ-170 Sentinel, and the RQ-4 Global Hawk—as well as human figures for scale (see figure 12). The project explicitly mimics the kits used in the training of gunners, radar operators, and civilians during World War II. During this time, American schoolchildren assembled visual guides by the thousands for civilian Skywatch groups, who monitored the skies for enemy aircraft. These collectives eventually formed the Ground Observation Corps, which at its height boasted 350,000 volunteers and over 16,000 observation posts and towers scattered throughout the United States. The circulation of these miniature planes among these groups provided a critical means of training subjects to identify unfriendly actors in the sky. In the process, it established a scopic regime of wartime security that sought to place the eye in the service of the creation and maintenance of the friend-enemy distinction, that critical opposition which Carl Schmitt understands as the origin of all "political actions and motives."[9] As Schmitt describes, this process of differentiation achieves its status not only by virtue of its connection to the real possibility of death, but also through its public and collective nature ("It is not *my* enemy, but *our* enemy"). In this, the identification models and the spotter industries they produced were bolstered by the logic of repetition, interchangeability, and seriality that defined midcentury industrial production. As the standardization of the latter allowed the eye to be trained and collectivized around national identity, the result

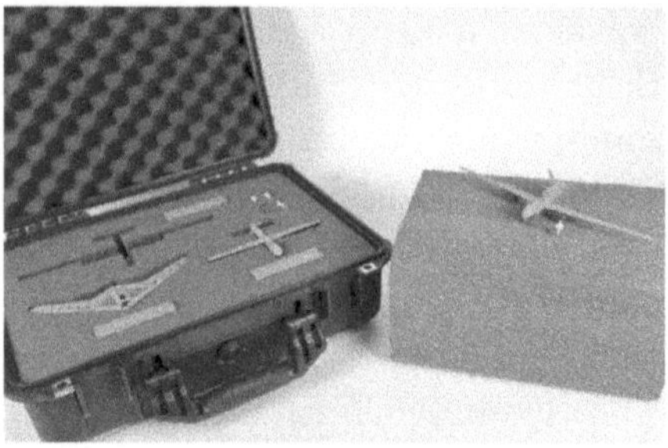

FIGURE 12. James Bridle, *UAV Identification Kit 001*, 2012. Courtesy of the artist.

was not simply that the enemy became recognizable, but also that a certain mode of processing visual information became formalized.

More recently, Galloway and Thacker have explored the possibility that the dynamics of the friend or foe opposition might be understood in architectonic terms. They point out: "Enmity is always 'faced' or constituted by a confrontation. We stand alongside our friends; I stand opposite my foe. Friends only 'face' each other insofar as they stand opposite and 'face' their common foe. . . . Enmity is an interface."[10] As such, Schmitt's basic unit of the political must be read as topological, i.e., as "mapped, superficial, structural and formal."[11] Interestingly, the historical evolution of these practices of identification suggests a similar formalization by material means. By the end of World War II, the centrality of the eye in practices of enemy recognition gave way to inorganic identification of friend or foe (IFF) systems. While these systems operate according to a variety of formats, the basic structure involves transmitting and receiving a pulse on a set of predetermined frequencies. As many have pointed out, such systems are structurally unable to definitively recognize an enemy as such. They can only confirm, with reasonable doubt, the presence of a "friend." As the inheritors of this technological bias, drones and the automated modes of vision that populate networked warfare regard everything as a potential target.

Bridle's *UAV Identification Kit* dramatizes the interpenetration of organic and inorganic agencies that undergird these relations of machinic vision by conjoining references to midcentury spotter industries with the medium of 3-D printing. Sacrificing the quantity of repetitive serial production for the uniqueness of micromanufacturing, 3-D printing follows a hacker logic, which eschews the notion of an authoritative original. In additive manufacturing, both files and objects are created and exchanged explicitly to be customized and remade. Circulating in the place of objects, stereolithography (STL) files present possible objects rather than durable goods. In turn, with the ability to transform objects into files via "rapid prototyping," the form of the object itself is virtually simultaneous with its future deformation. The medium is thus additive in a dual sense. Not only does it successively build up an object through the reaction of a binding agent with composite powder, but it also eschews negativity at a conceptual level.[12] After all, in this environment it no longer makes sense to talk about loss, authenticity, or even form in a static, eternal sense. Describing this condition in terms of a movement from objects to "the objectile," Deleuze explains:

> [T]his is a very modern conception of the technological object: it refers neither to the beginnings of the industrial era nor to the idea of the standard that still upheld a semblance of essence and imposed a law of constancy ("the object produced by and for the masses"), but to our current state of things, where fluctuation of the norm replaces the permanence of a law; where the object assumes a place in a continuum by variation; where industrial automation or serial machineries replace stamped forms. The new status of the object no longer refers its condition to a spatial mold—in other words, to a relation of form-matter—but to a temporal modulation that implies as much the beginnings of a continuous variation of matter as a continuous development of form.[13]

Shot through with this future-oriented ontology, Bridle's miniatures establish an explicit contrast to the World War II models and their correlative modes of industrial production. As the becoming of these thermoplastic objects undercuts their authority as guides, the work presents a practice of identification and targeting that is tethered to the network's relations of immanence. In this way, they reference a mode of violence that conceives of identity as a contingent category.

For the spectator, this conceit manifests as an unresolvable contradiction. On the one hand, the monochromatic surfaces, oversized carrying case, and excessively generic quality of these visual guides recast the drones as harmless toys. In this regard, one cannot help but read the dramatic shift in scale, tactility, and overall materialization of an otherwise unreachable entity as a gesture of mastery and containment. Yet at the same time, the groundlessness of 3-D printing undercuts these associations by shadowing the stability of forms with an unformed multiplicity, which evades such attempts at capture. When considering the recent engagement of military drones with 3-D printing, it becomes apparent that these tensions are not only nonantagonistic to drone logic, but in fact increasingly central to drone operation. In July 2016 the Russian defense firm United Instrument Manufacturing Company unveiled a printable drone that takes less than a day to print and around fifteen minutes to assemble. While this design utilizes additive manufacturing for quick on-site repairs or the mid-mission addition of drones to the fleet, the larger horizon of this interpenetration of drones with 3-D printing is seen in the recent projects of BAE Systems. The company is working on a new drone that is capable not only of repairing itself in flight, but also of breaking out into smaller, autonomous versions of itself while in the air. This pursuit of midflight shape-shifting can be traced back to 1999, when Lockheed presented its "small smart systems" to the US military. Chris Coker describes these drones as "having a wingspan of 6 inches, about the size of an outstretched hand. The object, which weighed only 3 ounces, . . . [is] able to fly in group formation or to detach into individual rotary-winged UAV's."[14]

Identification Kit draws a parallel between the tendency of additive manufacturing to produce multiple iterations from a single file or object and the ability of drone power to adapt by interpenetrating the singular and the multiple. While current research is attempting to manifest this capacity structurally, i.e., in the form of the drone itself, the larger dynamics of the signature strike and the operation of the kill chain embody these processes at a conceptual level. The conflicting relations that the spectator of Bridle's work encounters thus reproduce rather than resist the drone's own internalized tensions between ontological flux and static representation. By way of this enfolded opposition between representation and becoming, these plastic likenesses bear witness to the antiessentialist, relational aspect of networked warfare at the same time that they point to the persistence of hierarchies through which targets come into being as such. In this regard, the work utilizes twentieth-century references to issue a note of caution to the would-be drone hunter, as the stasis of representation—the guide by which one would seem to locate and contain the operations of power—unravels at the hands of additive manufacturing and the network that fuels its mutations. In many ways, it is a similar operation that is staged in the sonic field of drones by Ruben Pater and Gonçalo F. Cardoso's *A Study in 21st Century Drone Acoustics*.

DRONE SOUNDS, BIRD CALL

The booklet that accompanies Pater's collaboration with composer Gonçalo F. Cardoso contextualizes the drone within a premodern narrative of human-animal relations. It begins: "Our ancestors could tell a lot from looking up at the sky. Spotting and recognizing birds provided crucial information about the weather, where to find food, and if predators were near. In the twenty-first century landscape, knowledge of the natural environment has become our knowledge of technology. Most of us cannot tell the difference between the call of an osprey or a hawk, but we can tell the difference between a Samsung and an iPhone ringtone."[15] Citing the Federal Aviation Administration's (FAA's) prediction that there will be thirty thousand commercial and government drones in the sky by 2030, Pater and Cardoso go on to expand this analogy, claiming that just as humans once analyzed organic bird song to ensure survival, so must they now identify the sounds of "electronic birds" in order to deter surveillance and protect themselves from attack. To this end, the artists conceive of their album as a remake of the midcentury bird song records that honed the ears of listeners so that they might locate and identify these animals without recourse to the eye. But in the case of *A Study in 21st Century Drone Acoustics*, it is the identification of drones based on the sounds of their engines that is at stake. In order to better understand the strategic failures the work produces in regard to this endeavor, a brief discussion of the history and form of the bird song album is in order.

One of the most influential early birding albums is *American Bird Songs* (1942), a two-volume set produced by Arthur Allen and Peter Paul Kellogg for the Cornell Lab of Ornithology. The project began in 1929, when Allen was approached by the Fox-Case Corporation for help with recording birds using Movietone, a new synchronized motion picture sound system. In response, the Allen and Kellogg produced what is often cited as the first wild bird recording in a park in Ithaca, New York.[16] Despite being set in motion by the film industry, it was the outbreak of two world wars that fueled the evolution of these recordings. In fact, the genre of natural recordings of bird song was effectively born when the two men recast World War I parabolic molds as ornithological tools. Instead of picking up the sound of approaching enemy aircraft, these dishes now served to draw out bird song from the cacophony of the forest. Later, Allen and Kellogg were sent by the US Army to Panama to record sounds of the jungle, during World War II. The audio tracks they produced were played for servicemen while they slept so that they could acclimate to the jungle sounds before they arrived in Central America to protect the canal. While the confluence of military and civilian worlds that undergirded the *American Bird Song* album may have been buried when it was released to the general public, it was in fact the backing of the army that in many respects made its achievements possible.

As subsequent works, such as Donald J. Borror's *Common Bird Song*, Witherby's *Sound-Guide to British Birds*, and R. Tory Peterson's *Audio Field Guide to Bird Songs* attest, the Allen and Kellogg album is important not only for its entanglement of birding with military-based modes of sensing, but also for its establishment of what would become the standard format of the birding album. According to this formula, each track begins with a voice that announces the species of bird, sometimes including information about its appearance or behavior, before the call itself is heard. Occasionally the process is repeated for closer listening or to illustrate subtle variations between the species ("Listen again. . . . Hear that?"). Stripping away the layers of nonaural information that would comprise an actual experience of the bird in situ, the mobile recording process developed by Allen and Kellogg extracted a disembodied call from a sensory gestalt. Anthologizing these deterritorialized sonic events would therefore necessitate an intervention to restore presence and bestow identity. While supplementary literature and guidebooks included with these records provided important information, it was the human voice that was charged with transforming these otherwise illegible sounds into virtual presence.[17] This recurring act of linguistic enframing not only reaffirmed the coherency of the collection and its status as authoritative guide, but also provided the means by which a human narrator effectively delimited an avian subject. By training the ear to engage the natural world in these terms, the bird song album reinforced an anthropomorphic vision of the animal as both circumscribed by language and exceeding the symbolic systems by which it is given meaning. In this regard, the voice enacted a zoological variation of what Laplanche and Pontalis refer to as "foreclosure," a "primary form of repression . . . that is not performed by a subject but, rather, makes possible the formation of a subject."[18]

This convention of bird song albums using linguistic enframing as a means to re-encode sound that had been disembodied and decontextualized by the recording medium is critical for understanding the intervention of Pater and Cardoso's *A Study in 21st Century Drone Acoustics*. On each of the seventeen tracks on side A of the album, a deadpan, overly official voice announces a particular drone by name before the sound of its engines ring out for several seconds to a minute and then fade. Relative to its midcentury predecessors, which attempted to expand their audience by creating a soothing ambience that would attract birders and nonbirders alike, Pater and Cardoso's drone recordings provoke an undeniable sense of anxiety, exhaustion, and confusion on the part of the listener. This stems not only from the lack of nuance or variation of its rather relentless tracks, but also from the obliteration of contextual information or implied space. Unlike the lush, evocative sonic diegesis in which bird song tends to be captured (Is not part of the pleasure of these records the associations and images they produce?), these sounds of drone engines give little sense of recognizable or definitive environment. This sense of dislocation is exacerbated by what might be called the antimimetic quality of drone sound, i.e., its nonresemblance to itself.

Drones do not sound like aircraft. In fact, their underwhelming whir could not contrast more starkly to the gas turbine or piston engines that have occupied the skies for the last century. Inside of the Predator drone is a four-cycle, four-stroke Rotax 914 engine. As the anemic sound of its engines testifies, the propulsion system of this vehicle has more in common with the snowmobile than the airplane. Similarly, the Scan Eagle, a military drone used by multiple countries, features a two-stroke engine not unlike those found in chainsaws and weed eaters. By detaching drone sound from the seductive image of a sleek white body floating majestically in the clouds, these recordings excavate an alternate lineage in which the modern UAV is nearly unrecognizable. In this regard, what appears as homage to the bird song album is in fact an inversion of its basic premise. Instead of establishing an authoritative guide by which sound confers discrete identities, these modes of linguistic enframing dramatize a movement from mimetic sound toward what Deleuze and Guattari describe as a "generalized chromatism," in which "all components are placed in continuous variation."[19] As a result, the categories of identification that the narrator rather satirically proposes now serve not only to foreground the powers of deformation that undergird the operations of network-centric warfare, but also to actively unravel the singularity and coherence of the drone itself. Critical to this operation is the mode of production behind *A Study in 21st Century Drone Acoustics*, which replaces the mobile tape recorders of the original bird recordings with an even more deterritorialized mode of acquiring sonic data.

The drone sounds that comprise the album were taken from videos that had been uploaded to sites such as YouTube and Vimeo by a geographically and culturally disparate group of users, including servicepeople in the US and Israeli Armies, research contractors, and amateur drone enthusiasts. By outsourcing the recording process to the Internet, the portability of the recording medium that motivated the format of birding albums in the mid-twentieth century takes on a new dimension. Within this distributed mode of production, notions of authenticity, expertise, and authoritative representation prove unsustainable. As the work internalizes the rapid decontextualizing of user-generated content that animates these networks (the accompanying materials of the album meticulously chronicle the source and URL of each sound clip), the lingering presence of these concepts within Pater and Cardoso's anachronistic mash-ups only amplify the work's strategic inversions. As a result, the feeble attempt of the human voice to contain and delimit its subject by bestowing a discrete identity is overwhelmed by the deterritorializing capacity of the network. In this way, *A Study in 21st Century Drone Acoustics* translates the underlying relations of network-centric warfare into sonic form. In the process, a correlation is established by which the eradication of time and space that is enacted by the random access databases behind sites such as YouTube resonates with the disavowal of national boundaries and linear chronologies of the drone's eternal "everywhere war."

Both Bridle's *Identification Kit* and Pater and Cardoso's *A Study in 21st Century Drone Acoustics* integrate historical forms into digital networks in order to destabilize the systems of standards and protocols by which identity is formed. In terms of the former, it is the integrity of World War II "spotting" industries that is compromised by the endless circulation, abandonment of the original, and overall ontological instability of 3-D printing. Through this collision of historical referents and contemporary media platforms, the work dramatizes the underlying nihilism of drone warfare's eradication of time, space, and identity. Pater and Cardoso's album of drone sounds similarly compromises the solidity of reference by reproducing the form of the midcentury bird song album. By staging strategic failures within the taxonomic function of these sonic guides, the collection unravels the unique modes of presence that structure drone power. In distinct ways, both of these works seek to open ellipses in the symbolic register so as to subject the drone itself to the powers of deformation that undergird its operation.

THE MOBILE DEVICE AND THE NOW HERE OF DRONE DEATH

While the interventions of Bridle and Pater and Cardoso attempt to sabotage the asymmetry of drones by foregrounding the sensory experience of those on the ground, a second body of work takes this reversal one step further by interjecting these forces of deterritorialization directly into the everyday. At the center of this project is an effort to accelerate the tensions within the mobile device, to mobilize its dual identity as banal platform of civilian communication and medium of military aggression. As the investigation of the GeoCell database in chapter 1 illustrates, the cell phone acts as a primary means of targeting and surveillance in drone operation. Not only are the movements of individuals meticulously tracked by way of this device, but so are targets identified according to the patterns of communication it oversees. In several cases, attacks were even ordered based solely on the presence of a SIM card. Not surprisingly, many in Pakistan, Afghanistan, and Yemen closely correlate the mobile device with drone violence. In fact, in places such as Waziristan there is widespread belief that SIM cards emit signals to drones and that informants deposit US-provided cards at homes or businesses in order to signal specific targets for drone strikes. While the experience of the smartphone by citizens of the Global North may contrast starkly with these scenarios, the infrastructure that these "locative media" rely upon maintains close connections to the military networks that animate drone violence. Leveraging these layered symmetries, James Bridle and Josh Begley use the mobile device as a means to strategically unsettle the relations of invisibility and visibility that structure drone power.

In his *Dronestagram* project, James Bridle utilizes the GPS coordinates of drone strikes to fetch satellite images of the location, which are then captioned with information regarding the attack and posted to Instagram (see figure 13). As these posts provide seemingly live communication of drone strikes, the compulsory visibility that undergirds social media would appear to clash with the secrecy of drones.[20] And yet there is an undeniable sense that not only do the liveness and rapid accumulation of metadata of this medium provide oddly fertile ground for the presentation of network-centric warfare, but so does the urgency that structures these monetized communications reproduce the frenzied exchange of data that drone violence necessitates. This correlation is reinforced by the work's conflation of the vantage point of mapping satellites used by popular navigation programs with the camera eye of drones. Through this dual gaze, *Dronestagram* references a buried history in which many of the "satellite" views of Google Earth were in fact produced by drones, a relation that is only detectable by the occasional shadow that appears in the open fields and rooftops of Google Maps. (This phenomenon is discussed in more detail in chapter 2's discussion

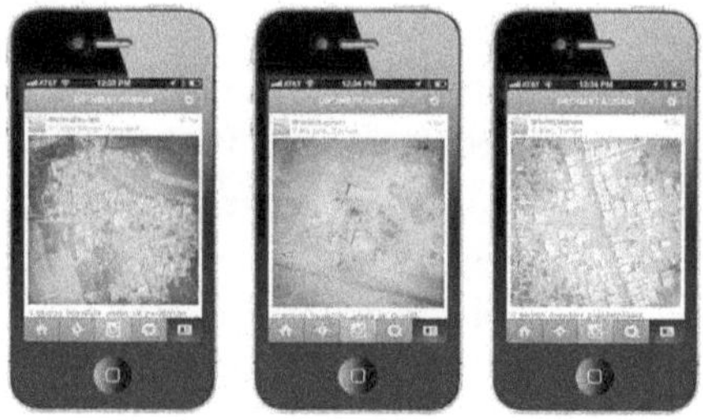

FIGURE 13. James Bridle, *Dronestagram*, 2012–2015. Courtesy of the artist.

of Bridle's *Drone Shadow* installations.) As the omniscient eye of the satellite's gaze is interrupted by the presence of the seeing body in these instances, a larger interdependence between the everyday spatiality of the mobile device and the martial networks of drone violence is revealed.

"Locative media" were inconceivable before the year 2000, when the full bandwidth of GPS satellites used by the military was opened to civilian use. Prior to this date, consumer technologies used the same array of satellites but were subjected to a system of "selected availability" (SA), which prioritized military operations and limited the bandwidth available for everyday uses. With these limitations removed, not only was the accuracy of everyday GPS increased tenfold, but so were the ties between locative media and this military infrastructure solidified.[21] As Bridle puts it, "these systems of visibility and surveillance [of social media] operate on the same substrate as the war machine, the drones themselves: networked communications, digital images, GPS systems."[22] This military-civilian fold flows in both directions, as not only is the spatial navigation of mobile devices thoroughly interpenetrated by drone logic and operation, but so is this network the very condition of possibility for the weaponized UAV. In short, the same GPS systems that inscribe locations on the shared photos of Facebook also provide the basis for what the Department of Defense website describes as the "precision guided munition strike" and "remote piloting of unmanned aerial vehicles."[23] In presenting inconceivable violence through a platform dedicated to selfies and food porn, *Dronestagram* exacerbates the contradictions and buried histories that animate social media networks. In the process, the everyday actions of social media users take on a meta-layer wherein martial networks and their civilian counterparts are interpenetrated.

In many ways, Bridle's project laid the groundwork for data artist Josh Begley's Metadata+ app (formerly Drone+), which sends a real-time message to users every time a drone kill occurs in one of America's undeclared wars. Begley describes the work in terms of an involuntary interruption, a "reaching into the pockets of U.S. smartphones users" in order to "annoy them into drone-consciousness."[24] In this regard, the app not only locates these attacks via GPS, but restages the event in terms of the breaks and discontinuities that structure the everyday experience of the mobile device. Despite being blocked by the Google App store for "objectionable content," the app is essentially a news aggregate, which draws its information from the publicly available database of the UK's Bureau of Investigative Journalism. Like Bridle's *Dronestagram* feed, its faux secrecy is strategic, as it insinuates a breech by which the co-constitutive relations previously described are rendered both legible and malleable. In this respect, what Begley describes as "drone consciousness" must be understood on two contradictory levels. Most immediately, it conveys the user's consciousness of drones, an awareness that unravels both the invisibility of drone violence and the asymmetries upon which its operation relies. However, it also speaks to the monotony or distracted state typically attributed to the mobile device—i.e., as the droning of consciousness itself. The strategic intertwining of these two trajectories is at the center of the aesthetic possibilities of locative media and their disruptive capacity.

The term *locative media* was adopted at the turn of the century in order to move away from the understanding of virtual space as an abstract absolute and to abandon the association of the digital with an eradication of embodied experience.[25] Presenting the experience of the user in terms of the "culturally specific and historically grounded concept" of place-making, this pivot sought to acknowledge the active spatialities of the mobile device and their concomitant social and political relations. Due to their ability to interject embodied relations and individualized practice into critical discourse, the categories of movement and mobility provided a crucial component of this challenge. After all, there would seem to be no better way to undercut the assumption that digital media "detach the viewer from an embodied, haptic sense of physical location and 'being-there'" than by insisting upon the primacy of the flesh and the localized productions of space these relations make possible.[26] However, it is here that the concept of "locative media" also reveals its limitations. As valuable as considerations of place-making and place attachment may be in disrupting early narratives of digital dystopia, these same methodologies can also overlook the ends-oriented logic and commercial interests that circumscribe experience.[27] In these instances, the user of the commercial platforms of the mobile device appears as a free and unfettered body whose "production of space" is disconnected from the monetized relations of surveillance it is inscribed within.

Rather than simply demythologizing or exposing these relations as false consciousness, mobile media art draws from relational aesthetics, performance art, and immersive theater in order to integrate this contradiction. For example, the work of artists such as Teri Rueb uses what she describes as the latent Cartesianism of "position-bound model[s] of space" as a means to create instances of "generative displacement."[28] In resurrecting the frictions that are typically "designed out" of commercial platforms, projects such as *Drift* (2004) and *Core Sample* (2007) prompt the user to "becom[e] sensitive to [the buried] thresholds of site and signal . . . [and] to feel a reconfigured relationship of the body in space as mediated by mobile technologies."[29] Echoing the accelerationist undertones of Begley's app, this work intertwines the forces of deterritorialization that animate the device (the drone of consciousness) with the capacity to produce reterritorializations (the consciousness of drones). In this way, what Mimi Sheller refers to as the "political paradox" of mobile media art—i.e., its immersion in "the dominant market-driven commercial promotions and state security driven surveillance"—becomes the basis for an accelerationist aesthetic in which the line between critical intervention and reproduction is necessarily blurred.[30]

Through these relations, the hybrid spaces of "locative media" are uncollated, producing impossible geographic and affective simultaneities that interrupt the ends-oriented nature of the space of commercial platforms.[31] Within these new distributions of affect, attention, and geography, the drone strike does not simply traverse spatial and temporal registers, but reconfigures the conditions of these categories. The experience of this event is no longer that of a unified subject traversing through continuous space, but rather a distributed mode of sensing that exemplifies what Ingrid Richardson describes as the "*openness* to the spatially non-coincident flesh of the world—not only bodily flesh but the 'stuff' of our environment."[32] In this way, the Metadata+ app corrals the asymmetries and potentialities of networks to nullify the elsewhere of drone death.

CONCLUSION: DRONES AS NONSECRET

Along with the familiar tropes of the "PlayStation war" and the "armchair soldier," the discourse of drone invisibility can conceal the violence that subjects who live in Afghanistan, Pakistan, Yemen, and other nearby countries experience on a regular basis. Universalizing the operator's gaze and the position of safety it articulates, such framings obscure not only the loss of life that drones bring about, but also the psychological torment they produce for those who live in their shadow. The work of Bridle, Pater and Cardoso, and Begley resurrects these experiences as a reciprocal gaze that strategically unsettles the tenuous position identity and geographic specificity maintain in these distributed networks. In so doing, it attempts to redistribute the asymmetries that martial networks articulate so that the elsewhere of drone power becomes present in the here and now. It is in this respect that the mid-century context that Bridle and Pater and Cardoso's work engages serves as an important foil to drone art's larger project.

Most immediately, these references to spotting industries serve to draw out tensions between the industrial production of the last century and the relations of becoming and objectility that undergird drone power. However, this historical mash-up also discloses a broader shift that drone art enacts with regard to the representation of disaster. Specifically, it dramatizes the way in which drone art not only rejects postwar conceptions of art, politics, and resistance, but also eludes enduring touchstones that have structured this discourse. Whereas postwar art registered the trauma of the Holocaust and the atomic bomb by emphasizing the impossibility of representation, drone art insists that embodied sensory experience be employed to unravel the war machine. Despite rejecting the use of symbolic ruptures to expose the failure of the signifier or the sheer impossibility of bearing witness, these interventions do not fall back on claims to realism or transparency. Nor do they attempt to convey some fundamental truth or brush with the absolute. Rather, drone art is purely productive, in that it seeks to push martial networks toward alternate iterations of themselves without necessarily articulating what particular form this reconfiguration might take. In this regard, the program of drone art is both nonutopian and nonmelancholic.

While drone art resonates with other twenty-first-century movements in its skepticism of ends-oriented practice and attendant dogmas, the genre's entanglement with the relations of invisibility discussed in this chapter must also be read as a symptom of a precise historical and technological moment in drone warfare. After all, it is the unique play between presence and absence produced by current drone operation that provides both the entry point for this work and the means by which it leverages the internal contradiction of martial networks. However, current research suggests that the window for such a strategy may be closing as drones increasingly jettison modes of explicit appearance in favor of a more absolutist approach to invisibility. For example, scientists at the University of California, San Diego, are developing a Teflon coating that will render drones virtually invisible by reflecting their surroundings.[33] British-based Plextek Consulting is working on electronic panels that use counterillumination to aid in the "visual signature suppression" of drones for those on the ground.[34] Attempts are also being made to eliminate or at least minimize the noise of drones using a system developed by NASA called Leading Edge Asynchronous Propeller (LEAP), which calibrates each engine to a slightly different RPM so as to spread the noise out across a broad spectrum of frequencies.[35] While these technologies have yet to be deployed by the military, they nonetheless suggest a discernible future of drone warfare in which this early dialectic between revelation and withdrawal is subsumed by an impending drive to invisibility.

In the meantime, however, drone power must answer a difficult question, namely, how to reconcile this structuring dialectic of presence and absence with the secrecy of its operations? After all, if strategic modes of revelation serve critical roles in the constitution of drone subjects, as the preceding discussion has suggested, one must wonder how it is that such instances of overt presence do not sabotage the clandestine nature of drone operation and/or contradict the recurring denials of drone violence by the US government. Pursuing these contradictory relations of the secrecy of drones not only sheds light on how the various works of this chapter achieve their disruptive power, but also helps to articulate the lure that drones maintain for the citizenry of the Global North.

Danny Weitzner, the director of the MIT Decentralized Information Group and host of numerous privacy workshops for White House officials, offers the following assessment to his clients: "[P]rotecting privacy in the digital age means creating rules that require governments and businesses to be transparent about how they use our information."[36] In other words, protecting privacy is not about ensuring restrictions on surveillance or encrypting information to block interception or prohibiting the blanket collection of personal data; rather, it is about rendering these interventions in the private sphere transparent. Implicit to this statement is the understanding of privacy as a particular mode of visibility in which strategic disclosure successfully oversees and manages the boundaries of adequate concealment. Weitzner's colleague Hal Abelson puts it this way: "[P]rivacy is the claim of individuals, groups, or institutions to determine when, how, and to what extent information about them is used by others in ways that affect them."[37] While there are obviously important differences between the confidentiality of everyday civilian communications and the classified operations of the US military, these relations nonetheless inform the contradictions that structure the official discourse that surrounds the drone program.

In an article aptly titled "How the Gov't Talks about a Drone Program It Won't Acknowledge Exists," Cora Currier describes the tendency of the Obama White House to extol the successes of drone warfare in the abstract while refusing to disclose specific details.[38] Under this strategy, high-profile kills are freely acknowledged, but the number of civilian casualties and the processes behind drone targeting remain hidden. While unremarkable in itself, this strategy of selective disclosure evolved into a contradictory knot of avowal and disavowal as the parameters of the drone program expanded. A pivotal moment came in 2011, when the American Civil Liberties Union (ACLU) brought a lawsuit against the US Department of Justice (DOJ) over a blocked freedom of information request concerning drone operations. The request and the subsequent lawsuit sought information concerning "the putative legal basis for carrying out targeted killings; any restrictions on those who may be targeted; any civilian casualties; any geographic limits on the program; the number of targeted killings that the agency has carried out."[39] CIA director Leon Panetta had repeatedly boasted in major news outlets about the success of the program from 2009 onward, and President Obama praised drone operations in press conferences and even referenced drone strikes on *The Tonight Show with Jay Leno*.[40] And yet the DOJ not only refused to provide the documents the ACLU requested in the lawsuit, it also refused to acknowledge whether or not the drone program itself exists.

The judge's subsequent dismissal of the case was based in part on a 1976 federal appeals court decision that had allowed the CIA to refuse to confirm or deny the existence of the ship *Glomar Explorer* in response to requests made under the Freedom of Information Act. In recent years the "*Glomar* defense" has been used successfully by a broad range of agencies for a variety of purposes, ranging from the New York Police Department's silence on the surveillance of mosques to the Department of Defense's (DoD's) unwillingness to release photos from the Abu Ghraib prison. While the argument has typically cited national security as its primary justification, the expansion of this doctrine under the war on terror has shifted the parameters of this defense beyond a purely

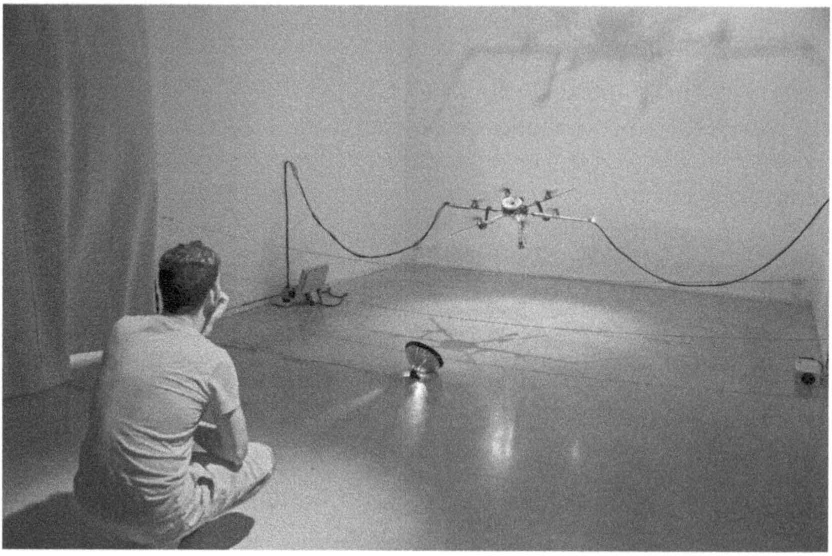

FIGURE 14. Nadav Assor, *Ophan* installation, 2014. A restrained hexacopter recites chapter one from the book of Ezekiel as it attempts to fly. Courtesy of the artist.

legal realm to everyday discourse. This expansion has largely been driven by processes of "securitization," a term introduced by the Copenhagen school of security studies to describe the way in which issues are strategically moved out of the political arena and into the realm of security in order to preclude debate or due process. Paradoxically, this process relies on strategic modes of transparency to establish the necessity of the secret. In this formation, sharing skeletal information or shards of intelligence serves to create the necessary conditions for the denial of information. In the case of drone discourse, these relations create a circular relationship wherein the active articulation of the nonexistence of the program validates its status. Far from undercutting the viability of drone power, the contradictions that abound from theses relations of securitization and secrecy reaffirm its extrajudicial status and justify its existence.

This movement between a policy of official nonexistence and strategic divulgence forms the basis of a larger operation of what I call "articulated secrecy," a recurring formation in discourse around drones that makes basic information public while withholding specifics and/or denying the existence of the program. These dual articulations of concealment and revelation disseminate "known unknowns," which utilize secrecy as a means to propagate specific modes of knowledge and experience. Former White House press secretary Robert Gibbs describes the difficulty of navigating these relations in the following terms: "When I went through the process of becoming press secretary, . . . one of the first thing[s] they told me was you're not even to acknowledge the drone program. You`re not even to discuss that it exists. . . . You`re being asked a question based on reporting of a program that exists, right? So you`re the official government spokesperson acting as if the entire program—pay no attention to the man behind the curtain."[41] As Gibbs points out, in the context of the administration's public communications on the drone program, the insistence upon the nonexistence of the drone program is not about concealing information. After all, in December 2011 Iran captured a US drone and disseminated images of the vehicle for all to see. In response, the DoD issued a statement confirming that it had lost control of one of its drones while flying a mission over Afghanistan, even while the *Glomar* defense persisted in official discourse. Such rhetorical strategies exemplify what Jacques Derrida calls the nonsecret.

For Derrida, the secret is "that in speech which is foreign to speech," in that its promise of a totalizing revelation is incompatible with the semiotic systems in which it is always already inscribed. On account of these relations, the secret operates via an "illusory depth" that conceals its status as "non-secret, . . . a semantic surface forever kidding us with a promise of depth."[42] As Ian Almond explains, for Derrida, "There are no secrets—or, if you like, there are only secrets, an endless succession of them, each one promising to be resolved by its successor. And this is precisely what a 'secret', traditionally understood, would be—a sign which would somehow, magically, unproblematically, explain all the previous signs leading up to it in one all-enlightening moment of magnificent self-presence. The disclosure of a secret would be the end of meaning. Nothing more could be said."[43] The secret promises an unveiling that language cannot perform. And yet it is precisely this shortfall, the gap between its promise and what its delivers, that provides the opening for a kind of secondary movement that preserves the "Big Secret," namely, that there is no secret (as Derrida puts it, "there is no secret as such; I deny it. And this is what I confide in secret to whomever allies himself to me. This is the secret of the alliance.").[44] In these terms, the secret of drone warfare is that its operation can only occur through the absolution of secrecy, an absolution defined both in terms of the voluntary disclosure of personal information that drone operation necessitates and the overt modes of disclosure that the drone apparatus itself performs in pursuit of its subjects.

On the one hand, the insistence upon the nonexistence of drones is a perpetuation of the illusion of the secret and the hierarchies of sovereignty that subtend its operation. On the other hand, the repeated acknowledgment of this nonexistent thing admits the true secret, that this ecology of the kill chain and the rhizomatic structure of the network-centric warfare disallow secrets. The "endless succession" of secrets that Almond alludes to is in this way more than a simple glut of noninformation propagated to conceal an absence. It forms a movement between conflicting utterances that materializes the Derridean nonsecret and in so doing initiates a self-sustaining loop of affirmation and negation, formation and deformation, which bolsters the fetish logic of drone power. Gibbs's reference to the climactic scene of the *Wizard of Oz* ("pay no attention to that man behind the curtain") is instructive in this regard, as it conveys the way in which the *Glomar*-ization of drone discourse shifts the register of information from knowledge to belief. Žižek has argued that it is ideological disidentification rather than traditional modes of interpolation that function to preserve power in the face of the cynicism of the current age.[45] In this context, the attitude of subjects toward authority often evinces the form of the fetish. By prompting what Freud explains as the processing of two "mutually incompatible assertions," drone discourse similarly reiterates the split that occurs within the subject at the hands of the fetish by way of the secret.[46]

This structure of the nonsecret runs throughout drone installation. It surfaces in the uneasy status of Paglen's "hacked" video footage of *Drone Vision* and in the presence of a concealed elsewhere that comprises Bridle's *Drone Shadows*. It also informs the unique relations of a work that has not yet been discussed, Jonathan Fletcher Moore's *Artificial Killing Machine* installation (2014; see figure 15). This "autonomous interactive mechanical installation" utilizes a publicly accessible database to mark drone strikes as they take place. When an attack occurs on the other side of the world, toy pistols fire and a small printer produces a receipt that contains the location, a brief description, and the minimum and maximum number of deaths produced by the attack. These receipts accumulate over time until an expansive ledger is formed, which accumulates into a pile on the ground below. Situated underneath the cap guns and small printer is a vacant chair for visitors to sit in. Two interrelated questions immediately present themselves in the face of this work: What subject position does this elusive chair desig-

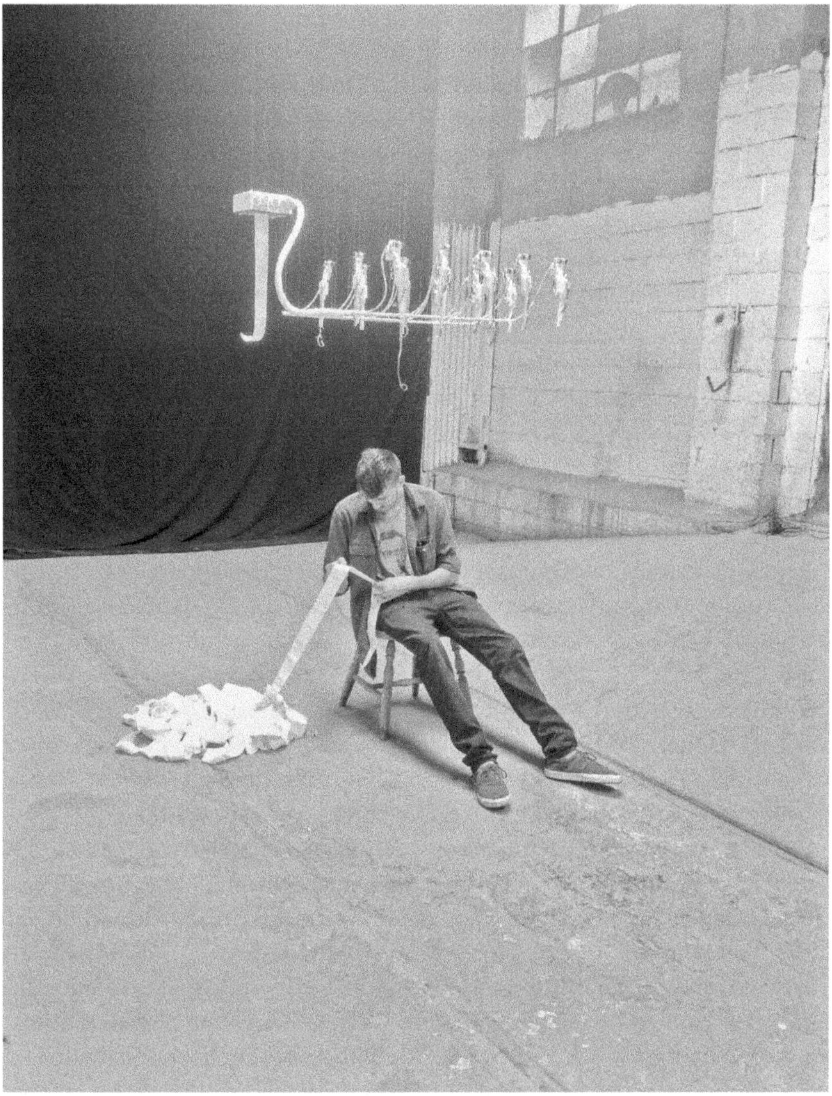

FIGURE 15. Jonathan Fletcher Moore, *Artificial Killing Machine*, 2014. Courtesy of the artist.

nate, and how can one reconcile the septic quality of this violence with the actual destruction drones produce? Tracing the complex web of fetishism and interpassivity that subtends these questions in Moore's work serves to reveal the conditions of possibility by which the nonsecret operates in the context of drone power.

The fifteen die-cast guns that hover from the ceiling in a perfect grid formation are based on molds of original toys produced by their last American manufacturer, Leslie Henry. Their quaint styling and nostalgic cap rolls seem to produce an absence of violence with each drone strike. Our only access to the strike would appear to be via the sculptural presence of these vintage toys, whose exaggerated stylings ironically conjure the face-to-face encounter of the ultimate symmetrical conflict, the duel. As the installation reproduces a larger process of removal from the strike through these relations, it forges an embodied experience that ironically precludes the actual sensations drones produce for those on the ground. In the process, this pile of receipts, fifteen toy guns, and an empty chair become strangely activated, offering the means by which the visitor is entangled with the founding elsewhere of drone warfare.

Through these relations, the apparatus of Moore's installation materializes the delegation that the kill chain enacts not simply among its operators, but also within the populace more generally. It does this by dictating an externalization of experience that proceeds not according to models of prosthetic extension, but by the logic of substitution ("for" rather than "as" us), in short by pushing the expectations of interactivity into the realm of what Robert Pfaller and Slavoj Žižek's describe as "interpassivity." While the term is often used by Pfaller as something of an antidote to the pervasive valorization of interactivity in the contemporary art world, its scope extends well beyond art and the contemporary sphere. *Interpassivity* references phenomena that do not simply contain in-built interpretations or "subject positions," but rather enact a displacement of agency or perception from the spectator to an object/device/person. Such works are "more than finished . . . [as] observers [are] relieved of the burden of creating as well as from observing [such that] the artwork . . . observes itself."[47] Žižek cites the "canned laughter" of TV sitcoms as an example. Here the program itself laughs for us, and the audience in turn experience this laughter as their own. Pfaller describes the larger ramifications of this instance in the following terms: "The TV comedy laughs about itself—in place of the observers who can delegate their amusement (and their 'passivity') to the art work."[48]

These instances of interpassivity often turn upon simulated representations, as in the case of the TV viewer who records a show without watching it, but nonetheless feels a kind of satisfaction by allowing the machine to watch for him, or the scholar who takes the book from its library shelf and photocopies the pages of the book and then goes "home with a deep sense of satisfaction—as if the machine had already read the text in [her] place."[49] In each of these cases, there is a pleasure in delegating to another agency, machine, or body. The interpassive subject is clearly aware of the fact that the act of photocopying a book is not the equivalent of actually reading it, but nonetheless the illusion persists through a kind of fetish logic that Pfaller associates with ritual. Such rituals have the capacity to displace action not only from the individual to the apparatus, but within the collective more generally. As Pfaller states: "The illusion at stake in the practices of interpassivity therefore has an interesting and peculiar kind of ownership: it is in a way nobody's illusion, an anonymous illusion, an illusion without a subject."[50] Through this constellation of interpassivity, fetishism, and ritual, the conflicting dynamics of Moore's work begin to come into focus, disclosing not only the conditions of possibility for the asymmetric war, but also the collectivizing tendencies of these relations.

Upon first encounter, one is tempted to read the empty chair of the installation as the negation of the subject that occurs via the remote relations of drone power. However, applying these relations of interpassivity to Moore's work suggests that the unique subject position the installation articulates is not simply that of an absent operator or an oblivious general public, but rather the bifurcated subject position of the fetish. After all, the sense of disconnect the work produces in relation to the event is simultaneous to the alternate modality of presence, which hinges on the temporal synchronicity between the experience of the installation and the violence it chronicles. As the firing of the toy guns and the printing of the receipts coincides with the launching of an actual missile on the other side of the world, this second order operation ensures that however spatially remote the strike may be within these mediated relations of the kill chain, it nonetheless occurs within a shared temporal interval. This sets up a relation wherein the viewer knows that he or she does not know, has not fully experienced, what has just happened, but nonetheless feels the sense of somehow having participated by virtue of this shared moment.[51]

By inscribing the violence of drones within these relations, *Killing Machine* performs the contradictions of the nonsecret as it migrates across the operational sphere of network-centric warfare and into public discourse. As the work couches this experience of drone power in the logic of substitution (a chair, a receipt printer, and fifteen guns perform the kill chain), it also dramatizes the way in which "known unknowns" structure a fetish logic that sustains drone power and its relations of articulated secrecy. While panopticism and its prioritizing of internalized modes of invisibility tend to elide these critical relations of presence, the works discussed in this chapter articulate more conflicted modes of presence in order to either forcibly interject the firsthand experience of drone subjects on the ground into public discourse or alternatively to weaponize the unreachability of these experiences for the citizens of the Global North.

CHAPTER 4

The Animal Remainder

Excavating Nonhuman Life from Contemporary Drones

Throughout history, war has served as a powerful catalyst for the colonization of the animal body.[1] Co-opting nonhuman life in order to extend the sensory and/or physical capacity of the war machine, these efforts have generally proceeded from enduring hierarchies between human and animal. The logic of the "war animal" thus appears to reiterate the modernist pursuit of an infinite expansion of human agency via prosthetic supplement, with the animal serving as simply one more "auxiliary organ" of Freud's "prosthetic God."[2] In practice, however, the entanglement of the animal body with the technologies of war suggests a less stable and deterministic trajectory, one in which the zoological component proves capable of not only exceeding these attempts at capture, but also destabilizing the anthropomorphic base of this narrative in the process. These seeming affirmations of human dominance over nonhuman life are in fact always already simultaneous with the production of what I call an "animal remainder." Most immediately, this remainder encompasses that part of animal agency that exceeds human influence at the site of "extension."[3] However, reading this excess diachronically presents the prospect of a second layer in this relationship, whereby a disembodied and virtual animal agency circulates within the machine, continuing to exert influence upon the apparatus long after the animal body has been left behind. Excavating this vestige of the zoological within contemporary martial networks not only presents alternate histories outside of the progressivist narrative of technology, but also opens up new ways of reading the various modes of convergence that accompany the trope of the animal in drone art.

This chapter begins the process of unearthing the animal from the modern drone by discussing the integration of birds into combat during the first half of the twentieth century. In these early experiments, birds not only offered a means of materializing an enduring connection between flight and surveillance, which had occupied the imaginary for centuries, but perhaps more important, introduced the time-space differential of an aerial proxy through which a thoroughly modern asymmetry of warfare would come into being. Two critical touchstones or "moments of rupture" animate this narrative: first, the displacement of motor and ocular functions to the animal via Dr. Julius Neubronner's experiments with pigeon photographers during World War I, and, second, the outsourcing of processes of targeting to pigeons in B. F. Skinner's Project Orcon. Through these early initiatives, birds were called upon to formalize a sensory motor schema that would echo through successive martial technologies before materializing in the technological interfaces of the modern drone. Pursuing this lineage suggests that not only do drones present a joint constellation that is animated as much by zoomorphism as anthropomorphism, but so does their operation conjure a lingering animal body that is both colonized by and determinative of its relation to the martial forces in which it is immersed.

This undead body provides a recurring entry point for drone art, which engages with the animal prehistory of the war machine by way of two inverted modes of convergence. The first seeks to challenge drone power by working toward interpenetration, enfolding, and hybridity, while the second pursues similar ends by actively extracting the human from its entanglement with the animal. Representative of the former is Mato Atom's *Seagulls* (2013), a short film that utilizes compositing, intertextuality, and nondiegetic sound to both destabilize human-centric histories of technology and recirculate the nonchronological animality that undergirds the everywhere war. Amplifying an ontological slippage among animal, human, and machine, the drone in this intervention comes to mark an atavistic convergence of histories and species, which exemplifies what Donna Haraway describes as a "colonial organism of machinic, human, and animal beings [born out of] historically situated infoldings."[4] An alternate approach, represented by the large-scale installation in Pakistan entitled *#NotABugSplat*, begins with the recognition that excavating this techno-zoological unconscious can very easily reinforce the very systems of power these gestures seek to reimagine. In fact, turning to the discourse of war and its tropes of animalized others suggests that these same relations also undergird practices of "bestialization," in which humans are associated with the animal in order to justify their status as targets. In response, an opposing set of artistic interventions comes into being, practices that invert the operations of Atom's film by utilizing processes of differentiation and individuation to wrestle identities free from the field of becoming that surrounds the figure of the animal.

By prompting the technological interface of the drone to divulge its animality, these works imagine not only alternate histories of war, but also alternate realizations of its technologies. In this regard, the inverse modes of becoming represented by the short film *Seagulls* and the *#NotABugSplat* installation act as complementary operations that collectively reproduce the animal assemblages of drone power so as to steer their capacity for becoming away from the imperial ends of networked warfare.

TOWARD AN ORNITHOLOGY OF DRONES

In the first decade of the twentieth century, a pharmacist in Kronberg, Germany, named Julius Neubronner began using pigeons to deliver prescriptions to a local hospital. After one of his pigeons lost its way and went missing for several weeks, the doctor and part-time inventor devised a way to track the routes of his birds. With lightweight, double-lensed cameras attached to their breasts, the birds would unknowingly take pictures of the ground below at regular intervals (see figure 16). The timing of this operation was governed by a pneumatic rubber ball, which was pumped with air beforehand according to the distance to be traveled. As the pressure decreased, a spring system triggered the shutter and advanced the film to the next frame.

Recognizing an affinity between the ability of these pigeons to grant visual access to otherwise inaccessible landscapes and the operations of military reconnaissance, Neubronner began a series of tests to gauge the readiness of these birds for the battlefield. The results were encouraging. Not only were the birds largely indifferent to explosions, but they were also skilled at flying low enough to get highly detailed images while at the same time evading enemy fire. The doctor quickly patented the invention under the title "Method of and Means for Taking Photographs of Landscapes from Above," and not long afterward the Prussian army purchased his cameras and flashy dovecote-darkroom combination used for devel-

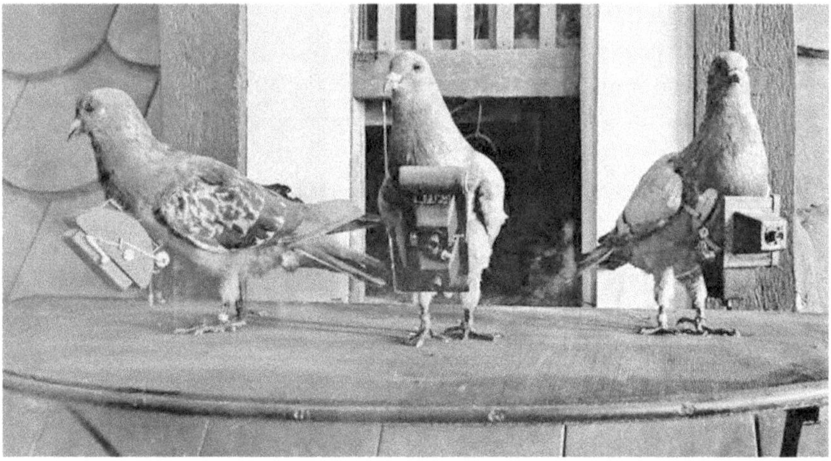

FIGURE 16. Dr. Julius Neubronner's pigeons, ca. 1910.

oping images on location.[5] As outlandish as this turn of events may sound to the contemporary reader, it is worth pointing out that pigeons have been a fixture on the battlefield for millennia, and the twentieth century was no exception. In fact, it is estimated that over a million pigeons were used during World War I as messengers. Successfully evading enemy fire, poisonous gas, and even trained hawks, the pigeon carriers' 90 percent success rate was the highest of any mode of communication used during the war, according to the Army Signal Corps.[6] However, the increased reliance on pigeon messengers cannot be solely attributed to the bird's skilled flying and hearty constitution. It was also the introduction of the camera to the battlefield that breathed new life into this ancient mode of communication. By allowing long messages to be photographically reduced to small payloads, the medium drastically increased the capacity of each "transmission," making for a much more efficient and sophisticated mode of communication relative to previous deployments of the "pigeon post."

While pigeons would remain an important means of transporting encrypted information throughout twentieth-century warfare, their stint as wartime photographers would be brief. Just as World War I was beginning, Neubronner was informed by the Prussian war office that "the introduction of aerial photography from aircraft" had rendered the project obsolete. Removing the parallax distortions and blurred focus that characterized the work of the avian photographer, airplane-mounted cameras produced standardized and scalable images. In this arrangement, gyroscopes, barometers, and the sensorium of human operators collaborated to consistently align the film plane with the horizon so that accurate, measurable distances could be extracted from aerial images. As a result, the coherent and singular vantage point of the photograph, which was subordinated to indices of animal movement in the pigeon photographs, was resurrected as a properly cartographic gaze in the aerial reconnaissance photographs that followed. In turn, the "modern aeroplane" was no longer simply a means for delivering bombs or transporting bodies, but now, as Lieutenant Colonel MacCleod put it in 1919, also served as "a traveling observation platform."[7]

Moholy-Nagy references this transformation in his essay "Directness of the Mind, Detours of Technology." Announcing the arrival of a "new vision" based on the objective operations of the camera, the publication juxtaposes images produced by Neubronner's pigeons with a composite or "mosaic" aerial photograph made by the airplane manufacturer, Junkers and Co.[8] The implication of the montage is clear: while the pigeon photographs undercut the anthropocentric ground of the image, their aerial successors reintroduced human reference in order to usher in the technical image. In so doing, the latter pushed the image in Hegelian fashion toward the realization of "Spirit." This familiar teleology, in which the animal is regarded as the means to propel the human sensorium beyond itself, reiterates a critical component of modernist aesthetics and its utopian aspirations. Summarizing this paradigm, artist Franz Marc asks on the eve of the war: "Is there a more mysterious idea for an artist than to imagine how nature is reflected in the eyes of an animal? How does a horse see the world, how does an eagle, a doe, or a dog? It is a poverty-stricken convention to place animals into landscapes as seen by men; instead, we should contemplate the soul of the animal to divine its way of sight."[9] As Marc believes that animals are "more beautiful, purer" than humans, "divining" their vision serves as the means of "giving birth to a new time," of "separat[ing] ourselves from the old world[.]"[10] In this relationship, the alterity that he ascribes to the nonhuman sensorium is in fact precisely the means by which the animal serves rather than escapes human vision.

Whereas modernism staged these interventions as "detours," momentary deferrals to the animal body in order to realize a new sensorium, the pigeon project in Kronberg articulated these same relations in less deterministic terms.[11] Specifically, as Neubronner's camera fails to penetrate the animal body in the way that Marc prescribes, it presents a mode of convergence that recognizes the persistence of nonhuman agency within these assemblages. Pigeons have a 340-degree field of vision, which is made possible by the bifoveal structure of their eyes. Unlike the human eye, which contains a single concentrated center of receptors that focuses vision in a frontal direction, the eyes of pigeons, like many birds, articulate two visual systems: frontal *and* lateral. This structural heterogeneity produces dual focal points, which do not necessarily operate in the service of parallax view, but rather allow vision to move between binocular frontal vision and monocular side view.[12] As Howard and Rogers describe: "When threatened . . . lateral-eyed animals, such as the pigeon . . . turn their eyes outward to gain panoramic vision at the expense of losing binocular overlap. When they wish to get a better view of something ahead of them they converge their eyes to increase binocular overlap at the expense of some loss of panoramic vision."[13] By organizing light around a single receptive surface of the film plane, the camera articulates a mode of seeing that contradicts the nonhierarchical relation between center and periphery that characterizes pigeon vision. The ascendancy of the camera's singular vantage point and comparatively narrow angle of view placed the pigeon in the service of a familiar anthropomorphic mode of extension.[14] To the extent that Neubronner's system offered a mode of seeing *from* rather than *as* the animal, it effectively prefigured the human operator that succeeded it. And yet despite this anthropomorphism, these photographs, which Ruth Erikson so wonderfully describes as "redolent with the activity of their own making," visualize an indexical presence of the animal that exceeds the confines of representation.[15] Through their off-centered compositions, expansive spaces, and hints of the animal body itself, the photographs summon an animal remainder, an excess that resides on the boundary of presence and absence.

In this tension between the human-centric translation of reality into an image and the indexical presence of the animal body, these images perform a kind of Benjaminian historiography, visualizing subterranean histories that destabilize familiar models of techno-determinism. They accomplish this by summoning an animal outside in futural terms as a presence that will circulate through successive technologies, first as a living body and then as a formalized mode of processing sensory information. This trajectory is confirmed by Skinner's World War II Project Orcon ("Organic Control"), a government-funded research project in which pigeons were placed in the nose cone of a missile so as to guide it to its target during flight (see figure 17). Inside, the birds would peck at an image of the target produced by a camera obscura. The locational information harvested by the screen would then be used to automati-

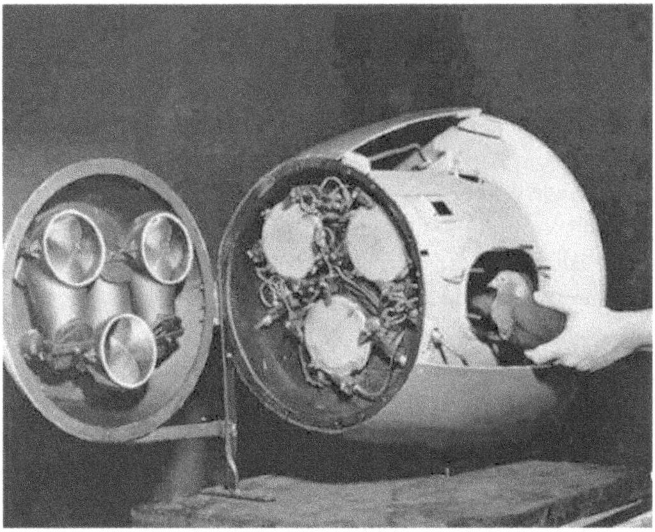

FIGURE 17. B.F. Skinner's World War II Project Orcon.

cally open and close chambers on the sides of the missile to guide its path.

Although Skinner's project proved highly effective in initial tests, it was eventually replaced by a guided missile system developed by RCA, which allowed human operators to remotely control missiles via a television signal transmitted from the nose of the explosive. Not only did the system use the same airframe as Skinner's project, but so too would the interface and screen of the pigeon be reproduced nearly identically for the human operator. In Skinner's prototype, the plate upon which the image was projected was coated with a semiconductor that was activated by a gold electrode attached to the bird's beak. RCA's "Bat" system revamped this technology as the "pick-off display converter," which allowed human radar operators to convey locational data by tapping a particular area of a screen with a special tool, a method that demonstrated greater accuracy and speed than verbal description. In Project Orcon, the lack of a common language between animals and humans necessitated an indexical and embodied mode of sign making. The zoosemiotics that emerged from this encounter would structure human interactions with remote technologies of war long after the disappearance of the animal operator.[16] One need only look to contemporary drone technologies such as ARGUS to find confirmation of this relationship. Here the avian prehistory formed by the work of Neubronner, Skinner, and others comprises an undead animal body that effectively redistributes the parameters of human and nonhuman agencies within martial networks in order to realize a comprehensive system of ground surveillance.

Focusing nearly four hundred sensors through multiple image-stabilized telescopic lenses, the ARGUS system is able to monitor all moving objects within an area of thirty-six square miles in real time from a single aerial vehicle. Live feeds are stitched together into an image, which acts less as a totality than as an interface through which users can access highly detailed, concurrent images alongside broader contextual views of the same scene and/or multiple fragmented views of other locations. The spatial montage formed by this live image mosaic actualizes an animal remainder of early systems, creating a hybrid mode of vision through which the human eye is able to take on an animal dimension. Freed from a singular position of observation and, more broadly, the directionality of perspectival media, "wide-area airborne surveillance systems" such as ARGUS testify to the presence of a disembodied animal, a presence that marked Neubronner's images as nonsymbolic residue or remainder, but which here enters the frame through the simultaneity of panoramic lateral vision and high-resolution frontal views. Similarly, the formalized gestures by which these modes of vision were made accessible to human operators in Skinner's project and the Bat systems that followed resurface in the familiar language of the touch screen. As the ARGUS interface translates taps, drags, pinches, and swipes to martial modes of surveillance, human fingers act as the beak of this missing animal body as they effectively reconfigure the image according to an avian mode of sensing.

This history illustrates what Donna Haraway describes as the coadaptation that follows from the convergence of media, animals, and humans. Writing on a spate of television shows that utilize "Crittercams" to give the audience access to the animal world, she explains:

> In so far as I (and my machines) use an animal, I am used by an animal (with its attached machine). I must adapt to the specific animals even as I work for years to learn to induce them to adapt to me and my artifacts in particular kinds of knowledge projects. Specific sorts of animals in specific ecologies and histories make me adapt to them even as their life doings become the meaning-making generator of my work. If those animals are wearing something of my making, our mutual but unidentical coadaptation will be different.[17]

While one cannot simply place a camera on an animal and expect to faithfully capture animal vision, Haraway reminds us that this does not necessarily mean that the medium of photography and its human operator are not reconfigured by virtue of this encounter with the nonhuman animal.

In this regard, her comments clearly go further than a simple critique of humanistic modes of extension via the animal. By associating this process with the extended temporal narrative of adaptation, she suggests that this relationship extends beyond the immediacy of these assemblages, forming a zoological paradigm through which histories of technology can be read. In these terms, the animal excess is not so much the causal agent behind contemporary drones; i.e., it does not provide the direct inspiration for or serve as predecessor to these technologies in a linear or deterministic way, but rather presents an outside that is partially recouped by successive generations of technology.

While discourses of automation and innovation obscure the animal prehistories that circulate within the modern drone, vestiges of this prehistory are sometimes made legible as these technologies migrate into civilian contexts. This is dramatized by the peculiar afterlife of Beatriz da Costa's *Pigeon Blog* project (2006–2008). Inspired by a photograph of Neubronner's camera-equipped pigeons, the artist attached air pollution sensors to pigeons in Southern California in the early twenty-first century. The data collected were then plotted in real time onto maps of the city and made available online. While using the animal to disclose the consequences of human intervention, the work was also intended to envision "new forms of co-habitation in [the face of] changing urban ecologies [via] a re-investigation of human to non-human animal relationships."[18] For this reason, the artist was shocked to receive an invitation to participate in a Defense Advanced Research Projects Agency (DARPA) grant geared toward the development of "small autonomous aerial vehicles" whose design was to be based on the biology of birds.[19] Perplexed, da Costa wondered: "How could *Pigeon Blog* possibly be of help to these people? Isn't it obvious from this work that a DARPA grant is the last thing that its authors would want to be involved in and that she is neither a biologist nor a veterinarian? Why was I suddenly being associated with areas of expertise that I was in no way qualified to respond to?"[20] While one can only speculate about what exactly motivated the military's interest in Da Costa's project, it seems reasonable to conclude that the work exposed an unrecuperated animal excess, a potentiality that held the capacity to both reinforce the ongoing anachronism between the technologies of war and realize alternate iterations of these relations. It is precisely this ability of civilian media to not only reveal this persistence of the animal within drone networks, but co-opt its powers of becoming, that the film *Seagulls* seeks to harness.

SEAGULLS: THE ANIMAL IN THE DRONE

Seagulls is a short, experimental work that centers on the relationship between a young woman and a drone. Throughout the work, the drone refuses the invisibility of conventional modes of surveillance and "speaks" to the unnamed female character through Facebook statuses and subtitles, which appear on the bottom of the film frame. Over a montage of the female character's birth, childhood, and adolescence, we read: "I watched you grow up. I followed you in the forest." At one point the young woman addresses the drone with a question that she types into her Facebook status window: "Why are your eyes so big?" As the film cuts to a close-up of the lenses within the drone's rotating turret, a subtitle explains: "So I can see you better." These enclosed and inaudible exchanges establish a sense of intimacy between inorganic surveiller and organic surveilled. Eventually, as the hierarchies of conventional modes of surveillance buckle under the weight of connectivity, a familiar horizontal mode of power comes into being. And yet despite the openness and transparency of these encounters, spatially *Seagulls* reiterates the remove that traditionally structures practices of surveillance. Communicating exclusively through images and screens, the drone and its subject occupy separate diegetic spaces throughout the film. Everyday scenes of the girl's adolescence are cross-cut with sequences of the drone moving through a kind of digital ether in which images of the same or similar events ebb and flow with the network's potentiality. Their paths flow in parallel to one another and never seem to touch.

All of this changes in the final scene, when the drone suddenly appears at a vacant beach where the young woman is sunbathing. For the first time this dispersed presence assumes a coherent and singular form, which hovers above the bikini-clad young woman. Without a word, it suggestively lowers its landing gear as if to penetrate her (see figure 18). As the young woman arches her body and quickens her breath, the plane abruptly turns into a blood-drenched wolf, frozen in pursuit of a red balloon. The use of the intra-frame transition of com-

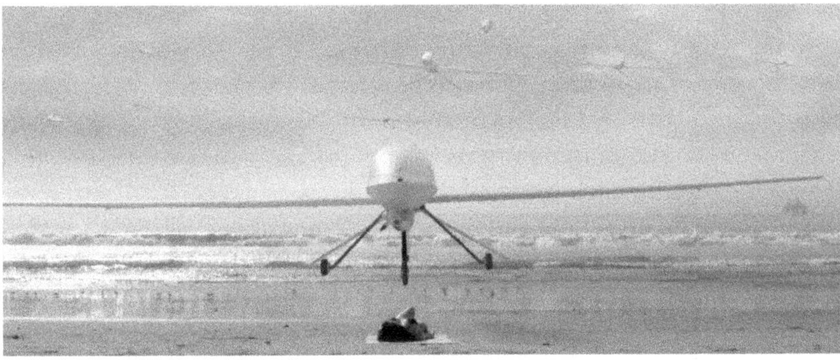

FIGURE 18. Mato Atom, *Seagulls*, 2013. Still.

positing positions the animal as the fulcrum by which the progressivist narrative of technology is folded in on itself, producing an archaeological history of the drone in which the animal is copresent with its inorganic successor. With the next cut the drone returns, only now it is surrounded by a swarm of identical UAVs floating in the distance. This panoramic shot diagrams the dynamic exchange that has taken place as the human in the foreground, the animal in the middle ground (represented by the seagulls), and the inorganic presence of the drone in the background now circulate within the frame as co-constituent forces. The interpenetration the scene visualizes is as much ontological as sexual, a shape-shifting that pushes this convergence of agencies beyond human-centric models of extension.

The ability of the film to divulge what Haraway calls the "animal-human-technology hermeneutic relation" is bolstered by its intertextual relations to the familiar fairy tale of "Little Red Riding Hood." As mentioned previously, not only does the unnamed female ask the drone, "Why are your eyes so big," but so does it admit to following her into the woods earlier in the film.[21] However, it is the abrupt appearance of the wolf at the end of the film that unlocks these otherwise buried references. The introduction of this intertextual layering at this particular moment is telling, as it suggests a productive overlap between the cultural function and historical roots of this tale and the animal-human hybridity that occupies the film in this final scene. In a well-known version of the story, Little Red Riding Hood is eaten and then cut out of the wolf's belly by a lumberjack, who then fills the animal's stomach to prevent him from running after them when he wakes up. Alan Dundes has argued that this feature of the narrative is a direct reference to ancient puberty rites, and that the young girl's dramatic emergence from the wolf is synonymous with her having achieved womanhood. In this, she moves from a state of nondifferentiation via this entanglement with the animal to a fully formed and autonomous individual who is recognized by the community as such.[22] In the case of *Seagulls*, this hybrid form turns back upon a technical infrastructure, associating the interpenetration of animal and human with the violence of drone power. As a result, the role of the sexual predator of the drone/wolf is to join with the human body so as to return it to a state of nondifferentiation that can be internalized and recirculated as potentiality within the technological interface.

The ontological instability that these processes rely on is reinforced by the sound track of the film. Composed entirely of lush synth music and other nondiegetic sonic data, the work's nonsynchronized soundscape undercuts what Michel Chion calls the "audio-visual contract" of narrative cinema, the implicit pact that ensures that audio and visual perceptions "mutually influence each other . . . lending each other their respective properties by contamination and projection."[23] The severing of this relationship becomes particularly significant when read in the context of the experience of drone operators. As numerous leaked videos attest, the images drone "pilots" and analysts work from are largely silent. As Nasser Hussain explains: "Although the pilots can hear ground commands, there is no microphone equivalent to the microscopic gaze of the drone's camera. This mute world of dumb figures moving about on a screen has particular consequences for how we experience the image. . . . [T]he lack of synchronic sound renders it a ghostly world in which the figures seem unalive, even before they are killed. The gaze hovers above in silence. The detachment that critics of drone operations worry about comes partially from the silence of the footage."[24] Conversely, for those living under drones, the experience of the UAV is closely connected to sound.[25] As discussed in the previous chapter, numerous firsthand accounts by Pakistanis confirm that they are typically made aware of their status as subject of the drone by sonic rather than visual means.[26] Thus, while pilots see but do not hear their targets, subjects on the ground often hear but do not see drones.[27] This "distribution of the sensible" creates a dispersed, nonlocalizable sonic presence on one side and a highly focused yet silent and nonreciprocal gaze on the other. By virtue of overseeing what is visible and invisible, audible and inaudible, the kill chain reinforces the asymmetry of drone power or what Ranciere calls the "police order."[28]

Seagulls internalizes this relation of drone warfare by presenting mute images whose only sonic data come in the form of an incessant synth-driven sound track. In freeing the image from the "audio-visual contract" of synchronized sound, audio information emanates from an unlocatable outside, summoning distant and flexible data registers that need not be tethered to embodied modes of witnessing. As this silenced world discourages conventional modes of cinematic spectatorship, which hinge on identification and embodiment, it creates the conditions for a mixed semiotics and its correlative aesthetic of accumulation. In this way, the film emulates the way in which, for example, live images are overlaid with data streams from SIM cards or audio recordings of harvested phone calls, forming parallel but co-constituent experiences that defy the singularity of place, time, or identity. With this web of shifting ontologies, the animal comes to loosen and transmogrify, forming an enfleshed convergence of forces that exemplifies what Akira Lippit refers to as an "animetaphor."[29]

Lippitt explains that as the animal is subject, without (human) language, the point of contact between the two systems marks the limits of the latter. And yet because of the "originary topography" that humans and animals share, the animal also brings something to language that is vital. Specifically, it presents the possibility of breaching the symbolic, enacting a specific mode of embodiment that Lippit describes as "a metaphor made flesh, a living metaphor [in which] . . . the metaphoricity of the metaphor collapses, [becoming] . . . metamorphic rather than metaphoric."[30] In its surpassing of the analogical aspect of metaphor, this constellation not only bears witness to a complex and nonlinear history between animal and human, but in so doing implicitly offers the means by which these assemblages might be opened up to their constitutive parts. As works such as *Seagulls* contextualize these relations within the infrastructure of the drone, this archive of technical transactions and animal sensing is rendered mobile and fluid, producing what Deleuze describes as "zones of indiscernibility."[31] In this way, the animal catalyzes a mode of convergence that resists both the humanist bias of "extension" and, more broadly, the hierarchies necessary to sustain the asymmetries of the global counterinsurgency.

With these relations in mind, the relative absence of the animal from the discourse of drone violence begins to make sense. As the presence of animality in the contemporary martial networks is obscured by the discourse of automation and its claims to have removed organic life from the site of conflict, the disruptive potential of this nonhuman agency is safely contained by familiar narratives of technological innovation. According to these progressivist histories, the technical evolution of the war machine proceeds by incrementally displacing "higher" forms of consciousness. From this perspective, the disappearance of the human from the cockpit is contingent upon an illusory disappearance of the animal from the battlefield that was enacted by the mechanization of war in the last century. The animal is therefore not only rendered nonpresent in the contemporary war machine in this formulation, but is in fact twice removed from the technical infrastructure of war relative to human agency. As *Seagulls* nullifies this narrative by interjecting the animal remainder of drone operation, it reveals the potentiality of an alternate framework. Through its use of compositing, intertextuality, and nondiegetic sound, the work not only discloses the ways in which animality serves as the condition of possibility for the everywhere war, but also mobilizes these relations so that alternate realizations of drone power become possible.

While it is tempting to universalize this strategy into a generalized program of opposition, the nature of drone power and the histories it incorporates refuse such a totalizing gesture. This becomes apparent when shifting these relations to the discourse of war and its recurring trope of the animalized other. Here processes of "becoming-animal" become entangled with practices of "beastialization," a shifting set of rhetorical and technical operations by which subjects are deprived of human status in order to justify their status as targets. As excavating the techno-zoological unconscious in this context risks reinforcing the very systems of violence artists seek to reimagine, an opposing set of interventions into drone power comes into focus. Inverting the operations of Atom's film, these practices utilize processes of differentiation and individuation to wrestle identities free from the field of becoming that surrounds the figure of the animal. In these instances, the strategic regression to stasis, specificity, and identity serves to interrupt a mode of martial becoming that itself utilizes the interpenetration of the human and animal as a means to engender violence.

WHEN HYBRIDITY BECOMES DEHUMANIZATION

Through his notion of "theoretical racism," Étienne Balibar illustrates the way in which a dialectical relationship between human identity and the animal reinforces racial violence. As capitalism demands that humans be "extracted" from animals by animal means (survival of the fittest, unfettered competition, and so on), the nonhuman is both internalized within the subject and posited as the absolute other. When this articulation of difference was paired with social Darwinism in the nineteenth century, it provided the condition of possibility for a self-sustaining racialized history: "Man's animality, animality within and against man—hence the systematic 'bestialization' of individuals and racialized human groups—is thus the means specific to theoretical racism for conceptualizing human historicity."[32] "Bestialized" at the hands of history, an identity of the other as animal circulates as "natural" within the present such that the indiscernibility of the human vis-à-vis the animal becomes the basis for repression and violence.

This regressive mode of hybridization is expressed most explicitly in the context of war, where establishing the animality of the other acts as a critical component of lifting the taboo against killing. Nick Haslam's seminal psychological study on dehumanization formalizes this relationship by identifying a specific "animalistic mode" at work within this discourse. In these instances, the association with the animal allows the other to be portrayed as devoid of self-control, higher level thinking, and morality, all of which serves to deny the subject his or her status as human life. These practices can be performative, as in, for example, the transportation of the Jews to concentration camps via overcrowded cattle cars or more recently the rituals of a group of rogue US servicemen who collected body parts of civilian kills as trophies as if they were wild game.[33] However, this animalistic mode is more often present at a rhetorical level, where associations of the enemy with the animal seem to function as a prerequisite for mass killings (the Nazis referred to Jews as rats, US soldiers described the Japanese as "mad dogs," the Hutus presented the Tutsis as cockroaches, and so on).[34] These historical relations of "bestialization" also inform drone warfare, which has in fact cultivated its own unique nonhuman rhetorical figure to expedite anonymous killings from afar.

A 2012 *Rolling Stones* article entitled "The Rise of the Killer Drones: How America Goes to War in Secret" revealed that it is accepted lingo among CIA, military, and drone pilots to refer to the casualties of drone strikes as "bug splats."[35] A charitable reading of this term might suggest that this dehumanizing of the target acts as an anecdote to the "near sight" that drone technology grants to operators. Such an interpretation would point to the apparent rise of post-traumatic stress disorder (PTSD) among UAV operators, which has resulted from the exceptional intimacy drones create both in terms of proximity (drone cameras can read a license plate from over two miles away) and temporality (unlike traditional bombing campaigns, it is not unusual for UAV pilots to observe their targets for days at a time). From this perspective, use of the term *bug splats* would seem to serve as a self-defense mechanism against this closeness, an attempt to compensate for the loss of distance by freeing operators from the messy obligations of intersubjectivity. This "Nuremberg reading" is undeniably tempting, as it preserves the "goodness" of individual actors, who appear to simply be caught in a technologically distorted and/or politically impossible situation. Unfortunately, it does not take long for the logic of this argument to begin to break down.

Most immediately, the term *bug splats* is born out of a similarity between the visual experience of shooting down a target on the grainy video feed of a drone's-eye camera and the crushing of an insect. However, this affinity itself is the symptom of a more primary intervention of simulation. In fact, the phrase was first used in reference to the virtual models of bombing raids produced by military software. When projected on large screens, the graphic indicating the site of detonation was said to resemble the remains of a flattened insect. In this way, the etymology of the term bears witness to an excessive mediation of the apparatus of drone warfare, a precession of screens, images, and data that conflicts with the excessive closeness that popular discourse ascribes to the experience of drone operators. Without going too far astray, it is worth briefly pointing out that the traumatic or even affective nature of drone operation is a hotly contested subject. The army itself has admitted to not having seen a single case of PTSD among drone operators, and one could feasibly argue that the discourse of virtual trauma serves an ideological as much as or even more than a medical purpose.[36] Specifically, by insisting upon the pilot's own psychological vulnerability as an adequate counterbalance to the existential vulnerability of the subjects on the other side of the screen, such a discourse mitigates the ethical problems that surround drone operation. Abandoning this psychoanalytic paradigm of trauma allows the analysis of *bug splat* to take into account not only the unique technical base of drone power, but also an intellectual history in which the image of the flattened insect corpse was regularly summoned to convey the anxieties surrounding modern vehicular travel and its interpenetration of speed, representation, and violence.

A variation of this trope appears in Paul Virilio's discussion of the violence of dromoscopy, a mode of vision that conceals velocity by flattening the world to the two-dimensionality of the screen.[37] Working from the phenomenology of driving an automobile, he describes a scopic regime in which objects appear to detach themselves from the horizon and move toward an immobile viewer. Conjoining the sensors of the dashboard with the controls of the vehicle (gas pedal, brake, steering wheel), the operator negotiates the images that appear on this screen and in so doing negates the extraordinary speed that makes this operation possible. Virilio explains:

> Arranged before the eyes of the driver this instrument panel forms a totality: the agent of displacement will by turns observe the approach of objects which will not fail to hit the windshield (images—but also *insects*, gravel, feathered creatures) and also diverse movements which will animate the gauges and counters. In this driving fascination begins a double game of lining up the inside and the outside of the car. With the help of the steering wheel and the accelerator pedal, the author-composer of the trip will in effect arrange a series of *speed pictures*, which will playfully sneak up on the transparent screen of the windshield.[38]

The bug splat, like birds and other organic materials that suddenly appear as flattened corpses on the driver's windshield, articulates a tenuous boundary between the internal space of the car and the passing landscape. In these instances, the otherwise inaccessible compressions of space and time that bring the outside into view as a technical image are momentarily made palpable for the operator. Theodor Adorno's earlier assessment of this scenario takes the correlation one step further, claiming that the automobile interface does not passively collect these animal corpses, but rather activates a bloodlust on the part of the driver. The "power of [the] engine," he claims, beckons the driver to "wipe out the vermin of the street" and thus produces a subject who has internalized this "fascist maltreatment" of the world.[39] In these formations, violence as the condition of possibility for the mastery of space through speed is made available via the animal body as it succumbs to the windshield/screen and its conceit of world as image. With this history in mind, it is only fitting that when Gregory S. McNeal attempts to communicate the blast pattern of a drone strike, his remarks correlate the dromoscopic experience of travel with the eradication of the animal body: "Bombs do not explode in a perfect circle but are flattened on one side, similar to the shape of a bug that hits a windshield."[40]

In industrial modes of transportation such as automobiles, planes, or trains, the bug splat resuscitates the otherwise inaccessible speed at which the operator traverses the landscape and in so doing foregrounds

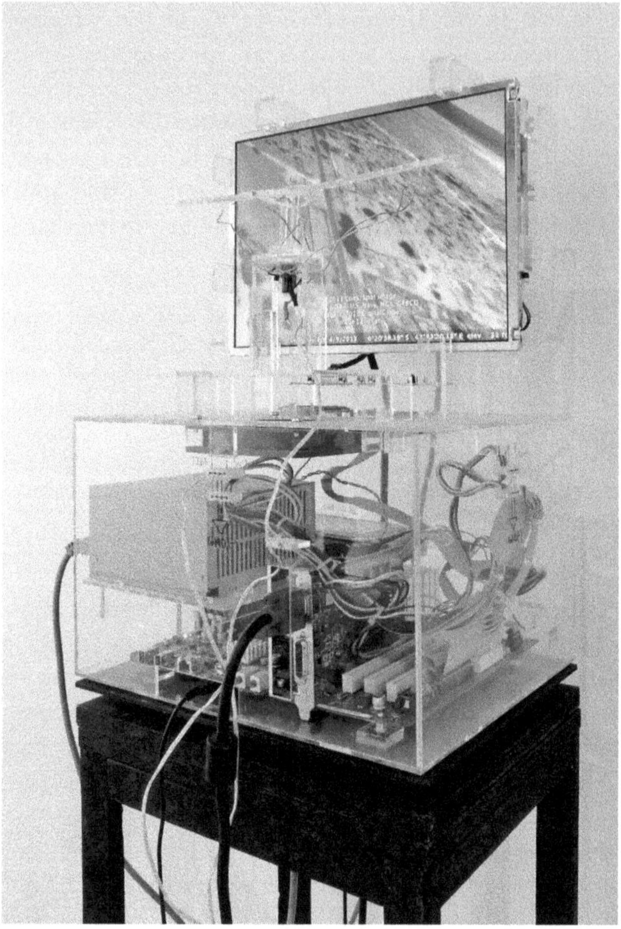

FIGURE 19. Lile Stephens, *Flight Simulator*, 2013. Courtesy of the artist.

the capacity for violence that these forms of locomotion carry with them. While drones are in many ways the inheritors of this trajectory of speed, representation, and violence, their operation extends beyond the dialectics of material and immaterial that structures these historical discourses of the "bug splat." This transformation is a result of a shift in the operator's relation to movement, which is dramatized by Lile Stephen's *Flight Simulator* (2014) installation. The work is comprised of a stationary plastic bird positioned in front of a screen on which a passing landscape is displayed (see figure 19). The stasis and overall artificiality of the bird, itself a stand-in for the simulated presence of the living agents behind the drone, contrasts starkly with the representation of flight on display. In fact, the ironic flashing red lights affixed to the wings and the diminutive size of the desktop screen give a sense of absurdity to the estranged relationship the viewer/operator maintains with the space in which the apparatus travels. Through these features, the work drives home not only the way in which the drone operator's experience of "movement" removes any danger from the experience of flight, but also the tendency of the screen to become informational rather than immersive as a result of these relations. The result is a purely virtual form of movement that travels through on-screen bodies and landscapes without the possibility of ever reaching the operator.

According to Virilio, the history of vehicular travel "progressively encloses the driver, first behind the screen of goggles, then behind the windshield, and finally in the interior driver's seat."[41] It would seem the inevitable endpoint of this incremental separation of the operator from "contact with the atmosphere" is the outright removal of his or her body from the vehicle performed by the drone.[42] However, in shifting the displacement of speed from imaginary to actual registers, the remote flight of the UAV enacts a kind of absolute transference of movement, in which the inadvertent appearance of speed is itself circumscribed by the virtual. As movement itself is rendered virtual by the remote nature of drone operation, the trope of the bug splat no longer interjects the absent sensation of speed and its attendant relations of violence, but rather opens to a larger set of symbolic relations that subtend its system of representation.

Read from this perspective, the apparent lack of specificity of "bug splat" is a symptom not of repression on the part of the operator, but rather of the conditions for networked warfare, which define targets according to a prepersonal, futural dimension rather than existing identities or affiliations. After all, a war waged without regard for countries, geographic boundaries, or finite objectives can only posit its enemy as an empty placeholder, a *possible* target that is defined by conditions rather than fixed identities. Not unlike the dynamic assemblage visualized in the concluding scene of *Seagulls*, the nonhuman animal here serves as catalyst and precondition for the entry of human subjects into this system. It is the mechanism by which the target is stripped of its subjectivity so as to be decomposed into metadata and fed back into the database as the condition for successive bug splats. The consequences of this relationship are perhaps most apparent when read in conjunction with the racial dimension of drone operation.

Instead of the overt racism that has historically informed wartime discourse, "bug splats" conveys the utterly generic, a condition more than an identifiable enemy, a situation or event instead of a bona fide "other." In this environment, even the "color" of targets would appear to no longer be a fixed quality and is instead relegated to the sphere of probability. As one intelligence official describes: "You . . . actually get a picture of a compound, and there will be something on it that looks like a bug splat actually with red, yellow, and green: with red being anybody in that spot is dead, yellow stands a chance of being wounded; green we expect no harm to come to individuals where there is green."[43] While skin color can never be wholly collapsed with race, the removal of this epidermal signifier by drone targeting systems nonetheless conceals the racial identity of the target. However, as Parks's reading of the drone's use of infrared illustrates, this concealment initiates a series of strategic contradictions that allow drone power to actually perpetuate racial hierarchies. On the one hand, by "turn[ing] all bodies into indistinct human morphologies that cannot be differentiated according to conventional visible light indicators [of] gender, race, or class," this mode of seeing understands all subjects as possible targets of drone violence. However, as this operation takes place almost exclusively within areas such as Pakistan, Yemen, or Somalia, it simultaneously preserves "existing power hierarchies" of race.[44] Reading the trope of the "bug splat" in combination with this (de-)coloring of the subject introduces a similar tension. Indeed, while the technological environment this term operates within eradicates identity, the term reiterates the "racio-speciesist" motivations of "beastialization" in its "reduction of bodies of colour to entomological waste."[45] Marking a summation of behaviors, practices and habits, the "bug splat" in this way serves as the technological equivalent of what Balibar describes as "racism without race."[46]

With all of this in mind, it is clear that utilizing the animal as an aesthetic strategy runs the risk of reproducing the kinds of "beastialization" that undergirds drone violence. *Seagulls* forged interspecies convergences in order to excavate an alternate historical narrative of the drone, one that displaced the progressivist, human-centric narrative of innovation with a more circular entanglement of human and nonhuman agencies. However, this program largely focused on the representation of drones themselves rather than their targets. The latter project requires an alternate engagement with the animality of drones, one that would understand the subject of drones in terms of a dialectical relationship to the relations of becoming that inhabit the kill chain. It is in this regard that the drone crash and the unique disruption it poses prove instructive.

As the crash gives concrete materiality to otherwise abstract and virtual relations, it draws out the tensions that animate drone operations in often spectacular fashion. As Jordan Crandall explains: "The geographical specificity of the material event [of the crash] reveals the distinct spatial politics in which these distributed drone economies are embroiled—economies that, in their vast scale and speed, and in their considerable rhetorical arsenals, blur impact at ground level."[47] Read from an accelerationist perspective, it seems plausible that this disruptive convergence of the distributed relations of network-centric warfare and the situatedness of this highly material event might be reverse engineered so as to provide the basis for an artistic intervention. In fact, this is precisely the premise of the *Drone Crash Incident* by artists Ricardo Dominguez, Ian Akan Paul, and Jane Stevens. Drawing from the genres of "critical fiction and disturbance theater," the group staged a drone crash in December 2012 on the campus of University of California, San Diego. After distributing numerous press releases, posting images on social media, and hanging posters on campus regarding the event and the ensuing investigation, a town hall meeting was held to discuss the occurrence with campus members and the local community. While the university subsequently denied the legitimacy of the crash as well as the existence of the Center for Drone Policy and Ethics that had organized the work, the collaborative performance nonetheless successfully harnessed the cathartic nature of the event in order to interject the invisible operations of drones into the public sphere. Not unlike the Iranian downing of a US drone discussed in chapter 2, the introduction of a particular community and place, what Crandall calls "geographical specificity of the material event," served to provide the conditions by which a collective reimagination of drone power could take place.

Extending this logic to the current conversation of animality, one wonders what would happen if the generic placeholders of the "bug splats" were made to appear with the kind of material specificity and discrete identities that the drone's system of targeting denies them. Could the processes of becoming that animate martial networks accommodate such heavily localized and stubbornly static signifiers, or would they buckle under the weight, finally delivering the asymmetries and space-time compressions of drone operation for a public for whom drones remain enigmatic and distant? It is precisely these questions that a recent art installation in Pakistan explores as it attempts to counteract the beastialization of drone violence. In interjecting particular places, singular identities, and individualized narratives into the distributed relations of drone power, the project effectively inverts the strategies of *Seagulls* so that the goal is no longer to engender strategic enfoldments between species, but rather to forcibly extract the human from the animal.

#*NOTABUGSPLAT* AND THE DISENTANGLEMENT OF THE HUMAN FROM THE ANIMAL

In the Khyber Pakhtunkhwa province of Pakistan, an artist collective installed a monumental image of the face of a child who had lost both her parents and two younger siblings as a result of a US drone strike.[48] Entitled #*NotABugSplat*, the haunting eyes of this ninety-by-sixty-foot portrait gaze up toward the sky as if addressing someone or something beyond the reach of human vision (see figure 20). According to the group, the intended audience for the piece is twofold. First, it addresses mapping satellites, which in capturing the image inadvertently create a "permanent memorial" to this young girl and her slain family that has persisted even after the work was dismantled.[49] Second, this face was intended to engage the eye of the drone so that the operator in Arizona and the martial network in which he or she was immersed would be forced to acknowledge this data artifact as a human subject rather than a bestialized other. As the artists put it, the oversized portrait "targets predator drone operators sitting thousands of miles away who refer to kills as bug splats. Now they'll see a child's face instead."[50]

Emmanuel Levinas argues that the face-to-face encounter provides the precognitive dimension of ethics. Critical to this status is the capacity of the face of the other to issue a particular and sustained injunction; it "orders and ordains" us not to kill.[51] It does so not simply by exposing the helplessness or "nakedness" of the other, although this is certainly part of it, but more fundamentally by summoning a universal law that calls upon the seer to suspend violence against this individual. For Levinas this "commandment" operates outside of symbolic relations and in fact precedes the will of the one who sees the face. Indeed, the nonviolence summoned by the face of the other cannot be attributed to human nature, nor can it even be fully explained by philosophy.[52] Far from the kind of "moral awakening" popular culture traditionally attributes to such moments of recognition, the ethics of the face in this

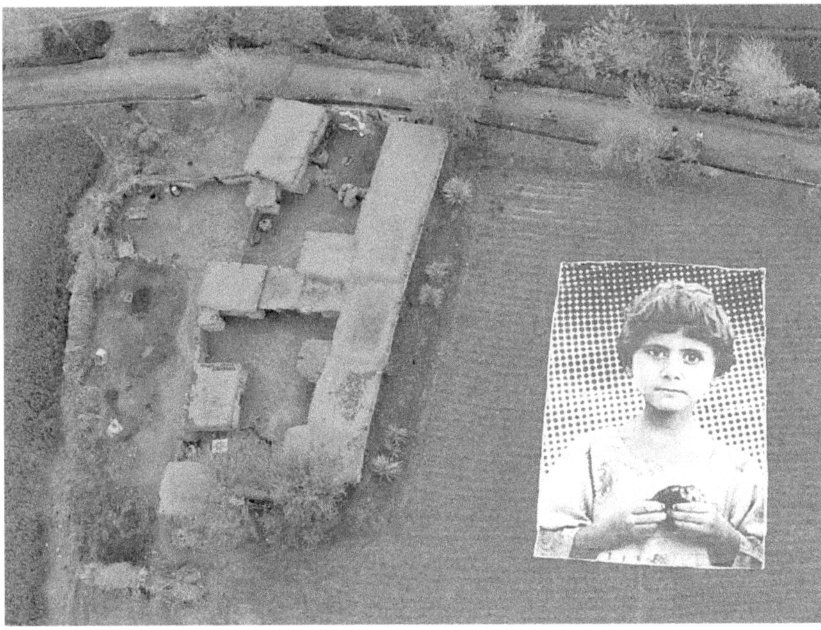

FIGURE 20. Ali Rez, Saks Afridi, Assam Khalid, Akash Goel, Insiya Syed, Noor Behram, Jamil Akhtar, and the JR Inside Out project, *#NotABugSplat* installation, 2014. Courtesy of the artists.

formulation is nondeliberative and nonreflective—hence its universality and power. As Levinas put it: "The first word of the face is the 'Thou shalt not kill.' It is an order. There is a commandment in the appearance of the face, as if a master spoke to me."[53]

In these terms, the appearance of the face in Khyber Pakhtunkhwa province would appear to be the ideal mechanism for confronting drone power. Shattering the narcissism of this martial gaze with a fully formed human "other," the installation negates not only the relations of bestialization that turn humans into targets under drones, but also the necessary absence of the face in these relations of violence. As Chamayou points out, "the resolution [of the images that members of the kill chain typically work from], although detailed enough to allow the operator to aim, is not good enough to distinguish faces. All that the operators can see are little figures blurred into facelessness."[54] This "facelessness" is representative of a strategic disassociation that Bräunert references in her insistence that "what the drone's cameras together with the lack of bandwidth in global data transmission achieve is . . . a highly functional condition that at least theoretically is meant to allow operators to fulfill their task without being troubled by empathy or relationality."[55] Following Levinas, the interjection of the face into these relations would then appear to act as a corrective to both the inability of drone power to recognize the other outside of the bestial mode and the excessive technological mediation upon which drone power relies, both of which strategically obscure the primary status of the face-to-face encounter and its insistence upon nonviolence.

While the negation communicated by the work's title certainly lends itself to such a reading, this interpretation fails to fully account for the multiple layers of mediation that occur at the site of the installation itself. Not only is the face itself a representation in the work, but so is the drone's eye that processes such an image heavily controlled by the inorganic agencies of the martial networks of which it is a part. Rather than being a simple likeness, this face marks the internal convergences of what Félix Guattari calls "mixed semiotics," in that it ropes together the computational aspects of the diagrammatic (the data of sensors, algorithms, and assemblages) with more familiar models of representation and/or expression (pictures, verbal and textual descriptions of the body). In light of these relations, it is not unreasonable to ask whether or not it is possible for the drone to fully "see" the human face in the sense that Levinas understands it. In other words, does this disentanglement of the human from the animal force drone power to acknowledge the ethical imperative to refuse murder, or does it inadvertently reaffirm the status of its target by presenting the face as always already an image? In engaging with these questions, an inverse reading of the *#NotABugSplat* installation comes into view, one in which the work does not so much interject the face as a means to summon this precognitive injunction against violence as seek a more primary ontological interruption by way of a refusal of the face and the politics of recognition that it engenders.

Rejecting the universal system of ethics that Levinas ascribes to the face, Deleuze and Guattari instead regard such encounters as enmeshed in a dual process of signification and subjectivation. As history confirms that such meetings are often synonymous with the imposition of a face onto the other, they argue that the face is just as capable of issuing a demand that the subject take on the face of the dominant order (first Jesus Christ and then the "white man") as it is of revealing the other as such.[56] Since the face is always interpreted and never a priori in this relationship, the processes of subjectivation that emanate from the encounter operate in tandem with systems of signification, which ensures the face only becomes legible as such within a dominant order of signs.[57] Recognition of the other is therefore not only nonuniversal due to its produced and situated nature, but also contingent upon the acquiescence to an authority on both a symbolic and existential level. This reading of the face is especially relevant to drone art's accelerationist tendencies and its skepticism of conventional modes of resistance. Consider, for example, Wendy Kozol's critique of drone activism's attempt to give a human face to these tragedies:

> Activists have sought to counter the imperial view from the sky by directing our gaze toward the ground, to pictures of casualties and destruction. Visual perspectives on drone warfare that bring the witness into intimate contact with human suffering use this imagery to refute government claims that few civilians have been killed in drone attacks. Intimacy, however, is no guarantee that the gaze can escape the historical racializations underpinning the War on Terror . . . putting a human face to an abstract discussion of "collateral damage" can reposition victims of drone attacks within a Western narrative of rescue. Moreover, technoaesthetic displays of U.S. military power visually compete for attention with these haunting figures of death. Ambivalences both within these images and as they operate intertextually thus complicate efforts at visual activism. . . . The "conundrum" is that visual witnessing as a process of witnessing operates relationally with dominant visual regime of U.S. imperialism.[58]

The contradictions of confronting drone power with the human face of its victims—its sentimentalizing of the other via narratives of rescue and structural affinity with the visual culture of automated warfare itself—call into question the ability of this image to humanize the other and/or undercut the violence the work references. It is for these reasons that, rather than sustaining a politics of recognition and its concomitant operations of territorialization, Deleuze and Guattari call for a "dismantling of the face" itself, an operation that requires the seer to "no longer look into the eyes, but to swim through them, to close [his or her] eyes, to close [his or her] own eyes."[59] According to this program, one must resist the very idea of a face in order to overcome its "despotic-authoritarian" nature.[60]

To better understand how *#NotABugSplat* engages with this alternate conception of the faciality, it is important to consider the fact that the installation presents a survivor of drone violence. While the deceased are more readily absorbed into the rhetoric of war, becoming "accidents," "targets," "tragedies," "martyrs," and the like, the survivor presents future-oriented consequences of death, which are less amenable to categorization and representation. We do not count survivors in order to measure the scale of the disaster. Nor do we typically erect monuments or memorials to them in the same way that we do for the deceased. The survivor is both interior to the disaster in the role as witness and yet denied its full dimensions on account of his or her survival. This liminal status is exemplified by the unique form of Maurice Blanchot's autobiographical tale, *The Instant of My Death*, which tells the story of a man who is "prevented from dying by death itself."[61] Most immediately, this conceit references an explosion that took place as the order was being given to a firing squad preparing to take Blanchot's life. However, this strategy is also a means to convey the contradictory status of the larger category of the survivor who, as Jean-Luc Nancy puts it, is not so much resurrected via this second chance at life as entangled in a perpetual operation of "resuscitating death."[62]

If this paradox of the survivor manifests in Blanchot's novel as "the story of a death which, properly speaking, the I cannot tell," then the installation in Pakistan bears witness to a similar bifurcation by way of its own poetics of the visual.[63] As a result of these interventions, the face the work presents is anonymous, nonpresent, already deceased, and yet undeniably persistent in its future orientation as survivor. Nameless and mute, she appears suspended in an image that robs her of her presence and yet she somehow survives the disaster. While the location of the work and its online presence are critical to this status, perhaps the most immediate contributing factor is the work's use of a child as its subject. As the unrealized potentiality of a child collides with the finality of death, the slippage of identity that the disaster provokes is further unsettled. Blanchot explains:

> "A child is being killed." This silent passive, this dead eternity to which a temporal form of life must be given in order that we might separate ourselves from it by a murder—this companion, but of no one, whom we seek to particularise as an absence, that we might live upon his banishment, desire with the desire he has not, and speak through and against the world he does not utter—nothing (neither knowledge nor un-knowledge) can designate him, even if the simplest of sentences seems, in four or five words, to divulge him (a child is being killed.)[64]

This death of a child, or rather the condition of a child being killed, enacts a multilayered transference across identity and time, a possibility that *#NotABugSplat* makes full use of. As a result of the work's active interpolation of drone networks, the subject's precarious position between life and death comes to operate between micro and macro instantiations such that this singular trauma repeats itself across national, racial, and generational lines. In this way, the absence of the child, which is both impending and already completed, issues a call or imperative by which the desires of this disappeared subject become our own, and in turn his or her now silent voice becomes the means by which we speak against the world that has killed the child. The work's insertion of the face into the field of drone vision therefore represents an attempt not simply to bring about a recognition of the target as a full-fledged other, but also to articulate the face as a dynamic absence so that the conditions of drone violence might be redistributed across the identities that this signifier typically circumscribes.

Critical to this capacity to dismantle the face is not only its subject, but also its location. The installation was originally intended to be placed in Dande Darpa Khel, the village in Northern Waziristan where the girl's family was killed. However, as the region proved too dangerous for outsiders, it was decided to locate the installation in Khyber Pakhtunkhwa, which has sustained significantly fewer drone strikes than its northern counterpart. Needless to say, critics were quick to point out that this location contradicts the artists' insistence that the work directly address drones themselves. However, placing the work in a part of the country that has comparatively little drone activity allows it to address a third audience, which exists outside of mapping satellites and the eye of the UAV. Summarizing the larger consequences of this move, Mike Pearl explains: "[T]he reality is that this isn't meant for drone pilots at all. It's meant for us, [the] internet."[65] While for Pearl this realization degrades the sincerity and political efficacy of the intervention, Tom Holert reads this shift as a kind of reflexive performance of the unique mode of sign production that informs drone targeting. According to Holert, this dislocation allows the work to "attain [. . .] an operability of a different kind [by] point[ing] to the fact that the visual and situational data on which drone operations are based purposefully exclude any face-to-face encounter with the casualties of contemporary warfare."[66] In short, by refusing direct contact with the drone's eye, the work dramatizes both the inability of drones to see the face and the necessary elision of identity that undergirds its processes of world-making.

This performance of "facelessness" culminates in the work's online afterlife. Here the digital image's prompt to labor—its call to share, comment, link, and like—produces a plurality of identities, contexts, and meaning, which weakens the photograph's claim to a unique likeness and, more broadly, its connection to the genre of portraiture. As the denotative aspect of the image is rendered contingent and fluid, these online representations come to act as a placeholder for a more generalized ontological movement rather than a definitive statement of the work's intentions or overall experience. The result, according to Kate Drazner Hoyt, is that the labor the work elicits from its "networked subjects" functions first and foremost as an interruption of the silence that structures public discourse on drone war. With the act of sharing taking precedence over any specific content, the political element of the installation's online presence is not necessarily ineffectual, but rather illegible as a coherent program of resistance. Instead, the digital artifacts of the piece oversee an elaborate web of displacements, which unsettles and redistributes the parameters of the conflict at a more formative level.

As important as the network is to this operation, this movement is inseparable from the unique mode of address articulated by the images themselves. Circulating aerial images of the work taken with a consumer drone, the project effectively conflates the viewing position of the drone operator with that of the social media user/viewer. By virtue of these strategic assemblages of text, media, and image, the act of circulating artifacts of the work on social media comes to expose the "material conditions that have allowed the United States drone war to be enacted in a relative dearth of public scrutiny."[67] For this reason, while some critics regarded the Internet-ready quality of the work as demoting what would otherwise be a serious intervention to the level of clickbait and "Internet heroin," Hoyt considers this shift in (or perhaps more accurately, expansion of) audience as providing the means by which the viewer is forced to "realize its own complicity in the attack."[68] In this relation, the labor of the civilian viewer/user effectively breaks open the face, dismantling the singularity with which it operates so as to take hold of its becoming. As a result, the face presents itself as an interface as much as an image, a prompt to accelerate the otherness of this image so as to redirect its powers of deformation.

While this intervention is staged in opposition to the bestialization of drone targets, the animal remains the necessary other for differentiation, the absolute not-me from which the human is born. Levinas himself experienced this interreliance while in the concentration camps, where he found that the only creature that would fully acknowledge the prisoners as humans was a stray dog named "Bobby" ("for [Bobby]—it was incontestable—we were men").[69] Perhaps then it is with a similar spirit of preserving the human through interspecies convergence that one should read the title of the work, *#NotABugSplat*. From this perspective, the apparent negation operates strategically, i.e., as the relations by which the animal makes the human possible once again. Just as the wolf is prompted to give up the young girl via the incision by the lumberjack in "Little Red Riding Hood," so does the intervention of *#NotABugSplat* force the human-animal assemblage to divulge the human so that she might be rescued from seemingly inevitable integration into the martial networks of drone power.

CONCLUSION: FLIGHT AND THE REVERSIBLE ANIMAL

John Durham Peters has described zoology as "the open book of media theory, in as much as it is a constant study of varieties of embodiment and the diversity of endowments."[70] As Peters's interest in the animal is part of a larger project of recouping natural formations such as fire, sky, and sea as instances of "elemental media," theory necessarily takes precedence in his project. However, this figure of the "open book" also serves as a productive means of thinking about media history, which itself is structured as a kind of bestiary through which one thumbs backward and forward in order to resuscitate various modes of embodiment and sensory experiences. In order to maintain the heterochronology of this field and its capacity for future enfoldings, media histories of the drone must necessarily internalize a series of contradictions between identity/hybridity, representation/becoming, and causal narratives/reversible archaeologies. As this history of war technology loses directionality and denies the singularity of agencies, it recovers the silences of the nonhuman animal that echo within inanimate machines.

This tenuous intersection of zoological, technological, and human spheres is in many ways the subject of a recent body of work by the Dutch artist Bart Jensen. The project began with the passing of Jensen's cat, who was aptly named Orville, after the Wright brother credited with ushering in the era of modern aviation. To pay tribute to his companion, Jensen created the Orvillecopter, a zoo-drone that resurrects his deceased pet as a working quadcopter. Echoing the surrealist strategy of "linking . . . two realities [together] that by all appearances have nothing to link them," Jensen's latest iteration of the series grants the ostrich, a usually flightless bird, a mode of aerial travel that befits its avian class.[71] The artist is also constructing a functioning submarine out of the taxidermied bodies of typically water-averse badgers, a work that is tentatively titled *Das Boot* after both the well-known film about World War II and the Dutch word for badger. While these pieces are shown (and flown) in art galleries, most viewers encounter them through the artist's rather unassuming YouTube videos, in which he stages test flights in what appears to be his backyard. Following a recurring format, Jensen's videos are bookended by takeoff and landing sequences that feature the artist manipulating the animal-drone via remote. The sense of control and intentionality that these shots establish is, however, quickly compromised as the animal elevates and takes to the sky, where the agencies of machine, animal, and human become less distinct. In this regard, these videos dramatize the works' capacity to perform drone art's contentious relationship to the animal, a relation which articulates both the symbolic rupture of interspecies enfoldment and the entrapment of anthropocentrism.

The distinctly human posture of these creatures, whose legs extend forward like the outstretched arms of a familiar gravity-defying action hero, is clearly the projection of an anthropoid fantasy of flight. This symbolic capture is made literal as these vehicles congeal into grounded sculptural objects. Resting on the pedestal in the white cube, the taxidermied flesh transforms these animal bodies into a kind of perverse trophy, conveying an absolute subordination to human will.[72] The act of flight, on the other hand, pushes these contradictions toward productive ends, revealing the co-constitutive nature of these agencies and the dynamic nonchronologies that subtend their histories. As these uncanny vehicles invite the viewer to confuse the locomotion of the machine with the bodily motion of this deceased animal, one has the impression that it is the undead nonhuman rather than the artist's manipulation of the remote control that leads the aircraft through space. In this way, the containment of the animal body gives way to unassimilability, an epistemological dissonance by which the animal escapes not only the functionality of the progressivist narrative of technology, but also the categories upon which such relations rely. Suspended between life and death via the absolute stasis of taxidermy and the uncanny locomotion of the UAV, the animal corpse performs the persistence of the nonhuman body within the war machine. As such, its movement is no longer simple locomotion, but rather the temporal and spatial performance of a counterrevolution between species and their technological supplements. This flight confirms the immortality of Neubronner's pigeon photographers, their silent insistence that the animal prehistory of drone technology not be sublated by its modern counterparts. It posits the animal as a structuring absence that haunts the drone's interfaces, operations, and possibilities.

As the UAV buckles under the weight of its new flesh, these works come to occupy the "zones of indiscernibility" that populate *Seagulls* and *#NotABugSplat*. In so doing, they reveal a shared necromancy by which the animal body is continually resuscitated by the violence of war. With war raging in Europe, the authors of *Royal Aero Club* magazine could not help but imagine the "compact stream-lined body" of the carrier pigeon as "an ideal base for the attachment of strong wings and highly flexible tail and for housing the powerful engine which supplies the energy for flight."[73] One finds more recent evidence of this necromancy in the description of a dolphin by a senior military strategist as a "self-propelled vehicle . . . with a built-in sonar system suitable for detecting and classifying targets; and carrying an onboard computer . . . capable of being programmed for complex performance."[74] And yet Jensen's zoo-drone refuse this recurring fantasy. After all, it is difficult to imagine how the inorganic appendages that adhere to this nonhuman body could establish tactical advantage. Hovering precariously in galleries and YouTube videos, these assemblages clearly do not assimilate the animal body as a means to improve flight, nor do they supplement the human eye by way of the animal-machine. This rejection of the prosthetic modality is evident not only in the sheer anti-utilitarian quality of these mash-ups, but perhaps more important, in the ambiguity they generate with regard to agency and identity. And yet their status as works of art, presentation as subject of the human gaze, and continual deferral to the human artist as origin of meaning all serve to return these bodies to the anthropocentric narrative of the wartime animal.

In this way, Jensen's zoo-drones embody drone art's conflicting approaches to interspecies convergence. Their movement between object and agency, stasis and flight, gallery and airspace charts the genre's contradictory program, in which the nonhuman animal functions as both medium for the interruption of drone power and means by which its operations are made possible. Whether enfolding species in order to unravel progressivist histories or extracting human identity from the animal to undercut their status as target, this group of works nonetheless operates from a shared desire to reproduce rather than resist the becoming of martial networks. Through this perverse symmetry, the animality of the drone becomes the mechanism for an accelerationist intervention, the means by which the disruptive power of network-centric war comes to serve as the basis for its own reinvention.

CHAPTER 5

Showing Sensing

Drone Space and Postmedia in Film and Theater

As the drone migrates from spaces of conflict to everyday visual culture, its dual status as weapon and medium creates the conditions for a unique mode of reflexivity. Disconnected from the scene of violence, the remote sensing of the kill chain manifests as a distributed mode of appearance in which the drone is never simply the subject of the image, but is always already the artifact of an absent martial network. In fact, its ability to achieve full presence on the screen, stage, or gallery is contingent upon an active interpenetration of the host media by the larger processes of network-centric warfare. The deterritorialization that ensues from these relations of intermediality encompasses not only the diegetic and/or institutional boundaries in which the work is presented, but also the medium from which the presentation takes place. To put this in terms more familiar to media studies, one might say that the appearance of the military drone in civilian media introduces specific modes of postmediality into the frame.

All "post-" designations are to varying degrees bound up with the bias of presentism, and the recurring diagnosis of the "postmedium" condition is no exception. While digital culture has certainly created new capacities for exchange between platforms, the collapse of seemingly infinite connectivity with the dissolution of discrete media that this term often conveys turns upon equal parts essentialism and technofetishism. As media are immanent to the organic and inorganic agencies, technical programs, and cultural practices that comprise them, their properties do not preexist individual iterations, but rather emerge from situated occurrences. Accordingly, the notion of a center or origin that current technology has somehow eradicated is not only illusory but ideological, in that it reinforces the underlying mythologies of a progressivist media history. For these reasons, Rosalind Krauss insists that "the specificity of a medium must be understood as differential, self-differing, . . . a layering of conventions [which] never simply collapse into the physicality of their support."[1] From this perspective, medium specificity is best understood as a paradoxical dynamic in which the singularity of the medium persists through its subversion, resurfacing as what Krauss calls "differential specificity."[2]

Focusing on the movie *Good Kill* (2014) and the theatrical production *Grounded* (2013), this chapter considers the ways in which the intermediality of drone warfare manifests in civilian platforms. As these two works tell extraordinarily similar stories through different media, they offer an ideal entry point for assessing the specific modes of media convergence that drones articulate within visual culture and the broader political and aesthetic consequences of these constellations. In addition, the specific media represented by this pairing, film and theater, are well-suited to this inquiry, as their mode of worlding has historically been defined in terms of their ability to assimilate other platforms. And yet, as this chapter confirms, the integration of the drone into these civilian forms is far from seamless. Despite their sleek photogenic surfaces, drones resist the role of passive object of a filmic gaze and instead insist upon being partial producers of a shared diegesis. As such, their presence maintains the capacity to interrupt enduring codes of cinematic identification by displacing the gaze from the invisible presence of the camera to an inorganic onscreen agency. What follows from these scenarios is an anti-immersive visual relation in which we no longer "see as" an onscreen subjectivity, but rather negotiate a series of onscreen "image objects" that effectively redistribute the image proper across traditionally noncinematic platforms. In these instances, the futural and data-driven nature of drone vision undercuts the medium's emphasis on resemblance just as its silent world threatens the audiovisual contract of commercial narrative cinema.

These conflicts are echoed at the level of plot, as the automated and remote nature of drone warfare clashes with the narrative conventions of the war film. After all, not only does the kill chain actively eradicate the notions of identity and individuality so privileged by Hollywood cinema (the lone soldier, the last man standing, the reluctant fighter) but so does the asymmetrical nature of drone warfare preclude the notions of risk that structure the genre's codes of heroism and suspense. (Is it possible for a soldier who sits in an air-conditioned storage container in Arizona to be a "war hero?") While all of this would seem to suggest that drone operation is incompatible with the medium of film, it is precisely these tensions that *Good Kill* cultivates in order to prompt the drone to divulge its inner logic. Through these conflicts, the work performs the ontological slippage of drone power, prompting the locations, narratives, and identities to lose their discrete quality and enter into a destabilizing web of simultaneity and connectedness. The recurring nostalgia for embodied modes of combat that frames this movement only further dramatizes the disruptive power of the drone's capacity for world-making. In this way, *Good Kill* ultimately illustrates the capacity of cinema to act as what Franco Berardi describes as conjunctive rather than connective media. While the former creates totalities that maintain the integrity of their parts, the latter redistributes the boundaries of individual identities in order to produce unforeseeable wholes.[3]

A similar relation informs the theater, which does not simply call upon external media to expand or deepen its diegesis. Rather, this integration of nonnative forms is constitutive of its mode of presentation, such that these "external" media effectively produce the category of theatricality with each production. However, this theatrical mode differs from its filmic iteration in the transparency with which it presents these relations. As conventional narrative cinema insists that the apparatus remain inaccessible to its viewers, the intermediality of the medium is often subjected to the same relations of disavowal that uphold the reality of the filmic image. Consequently, the appearance of noncinematic forms is often experienced as a conflict or tension in relation to an illusory essence. On the other hand, the apparatuses of theater need not disappear in order to fulfill their obligations to this diegetic world. In fact, as Chiel Kattenbelt describes, what makes the spectator's experience of theater unique is that it features not only a semiotic convergence of forms, but also a physical copresence of media. For these reasons, the possibility of reproducing the intermediality of drone power does not manifest as a conflict in the theatrical context, but rather as a shared mode of enfolding. This relationship is at the center of George Brant's *Grounded*. In foregrounding the congruity between drone warfare and the stage, the dramatic conceit of the play is entangled with the UAV's own processes of world-making. Along the way, the Aristotelian unities are sacrificed to what Francois Lyotard describes as a "general desemiotics" of theater, such that the same depersonalized relations of potentiality that animate drone power are employed by the play to delegitimize the narratives of insecurity and racial difference upon which its violence relies.[4]

While representing unique media environments and modes of convergence, both *Good Kill* and *Grounded* demonstrate a shared interest in integrating the creative capacity of the kill chain into their respective modes of presentation. With this gesture, they invite drone power to work upon the medium itself so as to reimagine the mode of critique by way of its object.

GOOD KILL: DRONE LABOR AND THE FILMIC DIEGESIS

Good Kill tells the story of Major Thomas Egan, a fighter pilot who served six tours in Iraq and now operates drones at Creech Air Force Base in Nevada. In his new position, Egan sits in a darkened storage container with other members of his crew, spending long hours surveilling suspicious persons as they go about their days on the other side of the globe. While occasionally featuring intimate moments and/or personal dramas, his shifts are generally comprised of a monotonous flow of banal events punctuated by rather anticlimactic missile strikes. Desperate to break free from the screens of simulated flight, Egan is willing to risk his comfortable suburban home life to get back into the cockpit of his beloved F-16. In fact, he even makes preliminary efforts to return to combat despite the misgivings of his commander and his wife. When these attempts fail and it becomes clear that he is bound to the darkened cockpits of Creech, Egan's longing for flight begins to give way to ethical misgivings regarding drone warfare.

As this nostalgia for the embodied sensation and affective intensity of modernist technologies of war pervades the film's spaces, the overall impression the viewer is given of the base is closer to an office park than a war zone. Figures haunt empty, white-walled rooms that are decked out in high-traffic carpet, folding event chairs, and motivational posters. The mess hall, that staple of the combat film so crucial to the camaraderie and fraternity of soldiers, is here replaced by a handful of sparsely populated picnic tables behind a chain link fence. Listless soldiers appear and disappear in these spaces, as if dazed by the abrupt transitions between reality and screen. Collectively, these familiar visual cues betray a larger conceit of the film, in which the eradication of sensation, embodiment, and presence that characterizes drone operation is communicated via the signifiers of postindustrial labor. This correlation extends beyond the spatial relations of the film. Subject to the on-demand temporality of global networks, drone operators and analysts maintain erratic schedules as they pursue enemies operating in vastly different time zones. Consequently, Egan rises at odd hours of the night and often arrives home as his neighbors are leaving for work and vice versa. Through these repeated rituals of disruption, the film collapses the instantaneity of network-centric warfare with the post-Fordist logic of the "just in time." As this labor erodes synchronized and regular schedules, the capacity of spaces such as the mess hall to forge a coherent collectivity is effectively undercut by this "always on" disposition of drone pilots.

These redistributions of time, space, and sociality are acutely manifest in the Conex boxes where drone operation takes place (see figure 21). As the embodiment of the monotony and alienation of earthbound aviation, these oversized cubicles quite literally negate the prospect of "actual" flight. Sitting on the runway, they form a physical barrier against flight and the excitations that come with this prospect. In the process, they also articulate an interchangeable architecture of beige that resonates with the rows of identical stucco homes in the Las Vegas suburb where Egan and the other drone pilots live. Emphasizing the modularity of these repurposed containers with shots of cascading sameness, the film foregrounds the intermodal capacity of these structures, correlating their insinuation of interchangeability and connection with the operations of network-centric warfare itself. In this way, the signifier of the Conex box pushes these recurring associations beyond simple homology and toward a more profound slippage in which the infinite variability of global supply chains comes to act as a stand-in for the modulated power of drone violence.

At the end of the twentieth century, Deleuze observed a waning centrality of disciplinary enclosures that delineate a unique set of laws and a subsequent rise of the dynamic molds of modulated space. As these spaces of the control society are animated by infinite variability, they demarcate what Deleuze describes as "self-deforming cast[s] that will continuously change from one moment to the other, . . . sieve[s] whose mesh will transmute from point to point."[5] In the context of *Good Kill*, this mutability is supplied by the immanence of the network, its ability to not simply commandeer time and space, but also to reconfigure identity within these parameters. At the hands of this disruptive force, loca-

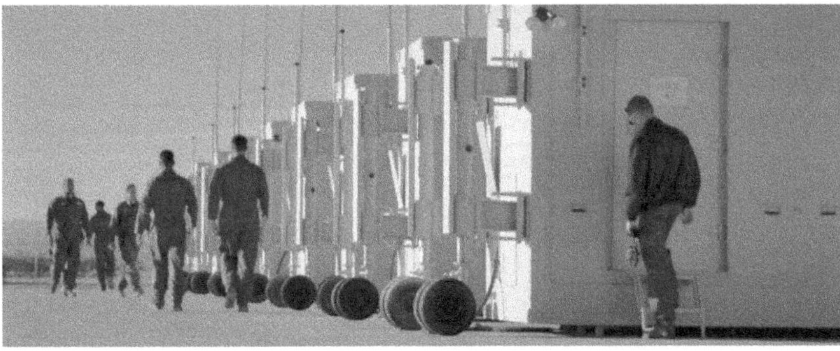

FIGURE 21. Row of Conex boxes, from the film *Good Kill*, 2014. Still.

tions not only internalize the movement between the factory, the barracks, and the home, so to speak, but do so according to a global program of strategic porosity, which is consolidated in the Conex box.

Sealed off to the outside, these storage containers block out the heat and light of the desert sun in order to allow powerful air conditioners to keep equipment from overheating. Just as Barthes recognized that the darkened theater served to strategically integrate the outside of the image into the erotics of the filmic experience, so do drone operators experience a breakdown of boundaries and fluidity of reference by way of the sealed space of "the box."[6] Its unique articulation of darkness creates an immersive environment through which a series of displacements of time, space, and identity take place (see figure 22). While these relations may center on the box, the firsthand testimony of drone operators confirms that these dislocations extend well beyond the base:

> We get out of our shifts—three 8-hour shifts—and the guys I was in a squadron with, we'd get out of the box and we'd go down to Buffalo Wild Wings, drink beer and debrief. It was surreal.
>
> When you are sitting in the box flying one of these things, you lose sense of the fact you are sitting in Nevada. You get so into what's going on. . . . You completely lose track of the fact that you are sitting in a Conex box, sitting in Las Vegas. You are pitch in the middle of this battle. . . . You're talking to the in-bounds; you're talking to the air controller; you're talking to the guy that's being shot at; you're trying to coordinate with the Pentagon; you're trying to coordinate with the combined operations center there in Qatar. Then all of a sudden the door opens on the box, and the next pilot and sensor operator walk in. And it's like, "Oh my god . . . I'm not in Afghanistan." It's bizarre—extremely strange.
>
> Normally, when you go to war, you go into a theater. . . .You sleep in a tent every night and you walk half a mile to go to the bathroom. In the

> Predator world, you're in Las Vegas. You get up in the morning, kiss the wife goodbye and drive up the base. But when you get into the box, you're right there in the theater. You're at war. It's incredibly strange.[7]

When writing the script, director Andrew Niccol interviewed multiple drone pilots in order to better understand the psychological demands of the position, and the sense of dislocation and strangeness voiced in these accounts figures prominently in the work. For example, when Egan stops at a local store after the end of his shift to pick up supplies for an outdoor barbeque, the man behind the counter points at his uniform and asks: "Is that real, I mean do you ever get to fly in a war or something?" Egan responds in a decidedly deadpan voice: "Yes, in fact I killed six Taliban in Pakistan this morning." The unsettling overlaps that drone warfare creates echo what Paolo Virno calls cognitive capitalism's "fluid relationship between labor and non-labor," such that the "sudden shift to different tasks, the necessity of adapting oneself to continuous shifting on-demand market" now describes the seamless movement between the virtual battlefield and everyday life.[8] As camera and characters move endlessly across the thresholds articulated by "the box," the repeated collisions of incompatible worlds eventually give way to impossible simultaneities of space and time. As a result, drone power manifests as a specific mode of porosity in which being at home or out with friends is invariably tinged by the experience of war. In turn, time spent in the cockpit is equally interpenetrated by the everyday outside, so much so that it hardly feels out of the ordinary when Egan casually texts his wife while executing a drone strike in Afghanistan.

In collapsing designations of inside and outside, military and civilian, here and there in order to produce the asymmetries of violence that comprise drone power, the film interrupts enduring conventions of the war film. At the risk of oversimplifying an expansive narrative, one might say that if post–Vietnam War films such as *The Deer Hunter* (1978) and *Born on the Fourth of July* (1989) refused enduring distinctions between the battlefield and the home front in order to draw out the psychological difficulties of returning to civilian life after combat, then *Good Kill* takes this process one step further, enacting a full-blown inversion of the dramatic codes of the genre so that it is war that is presented as uneventful and safe, while suburban life becomes uncertain and dangerous. At home, Egan attempts to hide his drinking problem, suspects his wife of having an affair, erupts in random outbursts, and struggles to transition between his day job as a drone pilot and everyday

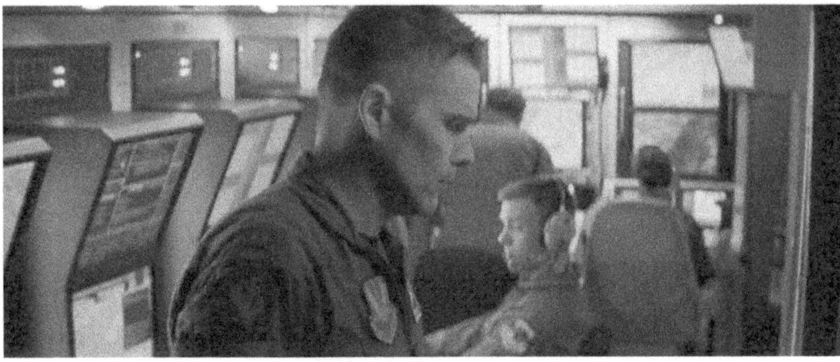

FIGURE 22. Interior of drone control center, from the film *Good Kill*, 2014. Still.

events such as the neighborhood cookout. During one particularly eventful night out on The Strip he barely escapes getting a DUI thanks to a sympathetic cop and attempts to negotiate the advances of a tipsy coworker. While civilian life is replete with drama and suspense, back at the base what would seem to be climactic plot points of conflict are repeatedly drained of anticipation. This is most pronounced in the kill sequences.

Each time Egan pulls the trigger to strike a target, there is an eight- to ten-second delay before the explosion appears on the screen because of the lag of the satellite uplink. Rather than condensing these intervals into tightly sequenced actions, the film revels in this elongated temporality. The endlessly repeated string of phrases that lead up to each kill ("Master arm on. Weapon's hot. Rifle. Missiles off the rail. Time of flight . . . eight seconds") only exacerbates the diegetic dead spots that follow. As the view of the drone camera moves into the film frame proper, decontextualized scenes of vast deserts thousands of miles away overtake the screen, presenting a kind of diegetic ellipsis or hiccup in which directionality, geographic specificity, and identity are confused.[9] Stripped of anticipation and drama, these kill sequences circumscribe a unique structure of affect, which might be described in terms of a particular mode of waiting. However, as the extraordinary asymmetry of this encounter presents the outcome as a foregone conclusion, the audience no longer waits for but rather holds expectation itself in abeyance. In this way, the film dramatizes what Garrett Stewart calls the "paradoxical terms of an unmanned heroics" in which "the very programming of a genre, even an anti-war genre, . . . seem[s] to have crashed . . . from the electronic overload of the plot."[10]

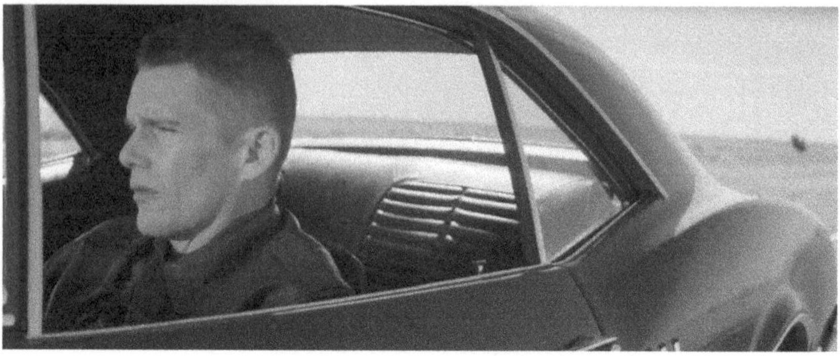

FIGURE 23. Interior of car, from the film *Good Kill*, 2014. Still.

This affective structure extends beyond the film's formal language and into the physicality of actor Ethan Hawke, whose slouching posture, nearly inaudible speech, and belabored movements read like a body shutting down, no longer engaged in the immediacy of the world. Describing Egan as "the action man whose freedom of movement has been snatched away," Anthony Lane suggests a usurping of the character's physical agency by a tortured mode of faciality: "Hawke, near-motionless for much of the film, registers the rise in pressure with his face. He purses his mouth, sucks in his cheeks, and drops his jaw, as if to show the skull beneath the skin. His eyes, beneath his crewcut, are raw and tight."[11] The intervention of the kill chain has not only severed the translation of perception into action, but also compromised the receptivity of the flesh, so that one has the sense that the hunched, atrophied physicality of Egan presents a body starved not simply for sensation, but for the very capacity for sensation. Appropriately, the only respite Egan can find from this state of corporeal deprivation is his *muscle* car, which dispenses with the dimensionless world of the technical image and once again renders the landscape as a traversable and embodied quantity complete with risk and freedom (see figure 23).[12]

A turning point occurs in the film when Egan's unit is tapped to work directly for the CIA. As this shift in command brings with it new tactics such as "signature strikes" and "double tapping," Egan begins to revolt against the idea of extrajudicial killings and the targeting of innocent bystanders who come to the aid of victims. Overcome by a growing resentment, he pretends to lose control of the drone during a particularly disturbing order to fire, giving his target just enough time to flee. After realizing his actions have been caught by his commanding officer, Egan locks the door behind his crew and in a climactic scene fires upon a civilian Afghani man whom he had seen rape a young woman on several occasions earlier in the film. Jettisoning the distributed agency of the kill chain, the scene reverts to a modernist mode of connectivity in which the apparatus functions as a prosthetic extension of an individual will. To this extent, it offers a kind of perverse restoration of Egan's identity as a pilot. However, the scene as a whole is highly ambiguous. After all, this attempt to restore meaning and valor through unauthorized acts of violence only reproduces the extrajudicial nature of this asymmetrical war. In order to draw out the larger implications of this sequence and *Good Kill*'s tenuous relationship to drone warfare more generally, a more thorough account of the film's entanglement with the reflexive dimension of the kill chain is in order.

Slippage, Symmetry, and Postmediality

In the course of reproducing the operations of drone warfare, *Good Kill* engages a complex network of gazes, which ropes together the computational aspects of the diagrammatic with more familiar modes of representation. While the Conex box may be encased in a veneer of banality at the level of narrative, the formal tensions that ensue from this "mixed semiotics" animate the "cockpit" with an ontological drama. In these spaces, the medium of film operates as a convergence point in which the organic members of the kill chain (pilot, image analyst, sensor operator, maintainer, and commanding officer) are imbricated with a series of nonhuman agencies. As a result of the drone's dual status as both object and origin of the gaze, these relations of the kill chain are interjected into the film's diegesis, initiating a unique mode of reflexivity by which the media ecology of drone operation both enters and occupies the frame.

This process first appears in the film's crosscutting between the outward stares of Egan and others in the control room with the views of the drone camera, then eventually extends beyond the confines of "the box," manifesting in a similarly disruptive use of aerial shots of Las Vegas and the surrounding area. After having spent multiple shifts in the cockpit with Egan and his crew scanning the deserts of Afghanistan and Yemen on pixelated screens, the viewer is unstuck when the film abruptly cuts to nearly identical shots of the Nevada desert. The slow cataloging of the landscape below, its transformation of space into a medium for targeting, collides dramatically with the domestic setting of these scenes. It would seem that the only thing missing from these shots of the American desert is the crosshairs of the drone's-eye view. These tensions escalate to a point where the film actually cuts between these two geographies in the same sequence, conjoining drone footage with close-ups of Egan and his wife gazing up into the sky. Without explicit narrative justification, these upward glances seem to briefly acknowledge the kill chain's collapse of time and space, before the diegesis congeals again and the characters return to watering the lawn or cutting the grass. As this geographic slippage comes to include relations of identity and affiliation, a peculiar symmetry takes hold of the plot, a congruence by which the lives of the drone operators and analysts seem to form mirror images of those that occupy their screens each day.

Members of the kill chain watch children playing soccer, women doing laundry, and spouses arguing during their shifts at Creech and then return home to effectively reproduce the same events within the context of their suburban lives. These dynamics culminate in a rather peculiar scene in which Egan suddenly finds himself parked in front of a mosque in Las Vegas, staring through his windshield at a group of Muslims as they leave the building. Without setup or explanation, the scene appears abruptly. It is as if Egan (and by extension the film itself) has confused and/or collapsed multiple entities: the drone and the automobile, the screen and the windshield, the Arizona desert and the arid landscape of Afghanistan, Muslim Americans and terrorists. Despite posing something of a non sequitur at the level of narrative, conceptually the scene succinctly dramatizes the way in which drone vision interrupts not only spatial and geographic continuity, but also discrete identities. In their place, it presents a series of suspended binaries through which the projective function of the signature strikes takes place.

In taking stock of the larger consequences of these interventions for the film's message, one quickly runs into the paradox of drone cinema. As drones act as both medium and image, their onscreen presence inevitably integrates the media ecology of the kill chain into the creation and organization of the cinematic diegesis. Seeing as though this assimilation prompts the filmic frame to engage with a mode of sensing that exceeds purely symbolic relations, the camera's acquiesce to the drone's eye has the effect of interrupting the cinema's reliance upon on iconicity and representation. As such, not only do drone "texts" often sacrifice the ability to remain external to their subject in the course of integrating these systems into the frame, but so can their specific form complicate the kind of meta-level critique typically associated with the antiwar film. *Good Kill* exemplifies this scenario. Despite the expectation among critics that the work represent the first antidrone movie, the disruptive potentiality of the film is safely contained between two opposing narratives. On the one hand, its mirroring effect humanizes the other and thereby implicitly critiques the exceptionally elastic boundaries of the category of "the enemy combatant" that undergirds drone warfare. How, these symmetries implicitly ask, can these subjects possibly be such dangerous threats when their lives so closely resemble "ours?"[13] On the other hand, these same relations can just as easily be read as reaffirming the insecurity, vulnerability, and fear that undergird the global counterinsurgency. From this latter perspective, the film's integration of these spaces into the domestic everyday materializes a familiar post-9/11 discourse of terror in which large-scale, heavily orchestrated attacks give way to the "homegrown terror" of "sleeper cells," small groups, or even "lone wolves" who remain virtually undetectable due to their full-scale immersion into the dominant culture. To this extent, the film's processes of enfolding prompt a familiar anxiety regarding the eternal nearness of terror, a "structure of feeling" that has historically operated in the service of the global counterinsurgency.

The inability or unwillingness of *Good Kill* to reconcile these readings, let alone answer the questions it raises, is no doubt partially to blame for the lack of enthusiasm that critics seem to have for the film.[14] While the film sets up the expectation of a condemnation by way of the ethical misgivings of its protagonist and overall demythologizing of drone warfare, the final statement is one of ambiguity and even lack of coherence. François Truffaut famously commented that there is no such thing as an antiwar film, and it is tempting to consider this apparent shortcoming of the film in these terms, i.e., as a symptom of the medium of cinema, which is itself predisposed toward spectacle and the glorification of combat.[15] However, writing in *Film Comment* in 2009, Garret Stewart describes the status of the war film after drones in terms of an opposing problem:

> American films about Iraq or Afghanistan keep failing at the box office. We're told in the press it's too close for comfort, too soon for analysis, too much like free TV to warrant ticket prices. The real problem is that it's too

> shapeless for plot. . . . [It lacks] a formal coherence taut enough to secure a cogent oppositional stance. Instead, the frequent anti-war *burden* of these films feels like stress only in the other sense, disorienting and demoralized—especially when the most salient formal feature [is] the blanketing of plot by a thick quilt of digital mediation, . . . the high tech fantasy of an ultimately unmanned victory. . . . All of it [is] saturated by video: not only second-hand embedded footage but the ubiquitous remote control of drone predator surveillance in the high- tech fantasy of an ultimately unmanned victory.
>
> [T]he damage is inflicted not just on characters but on plot. Narrative agency is subsumed to technology at every level, from aerial tracking, where characters are just faceless pawns on a monitoring grid, to eye-level confrontations, where any human posture toward an encroaching violence, from suspense to panic, often feels as virtual, as permeated by mediation, as computer interactivity in some low-resolution videogame. For the soldier to see action in this context is more like an ocular than a military idiom.[16]

Positing the presence of drones and digital media as an intrusion that corrupts properly cinematic operations inevitably frames the twenty-first-century identity of the war film in terms of an existential crisis. Such diagnoses are reflective of the larger paradigm of "postcinema," which, as Steven Shaviro describes, understands the historical role of film as a "cultural dominant" diminishing at the hands of a series of competing media platforms that have arisen over the last half century. Following this trajectory, one is left to wonder, as Stewart does, whether the genre of the war film (and perhaps even the medium of film itself) can survive the advent of automated combat.

A more nuanced reading understands the drama of the frame that typifies these films in terms of what the art historian Rosalind Krauss refers to as the medium's "differential specificity," its inbuilt tendencies toward "self-differing." As D. N. Rodowick argues, since the arrival of sound, the "infringement" of television, and even the enigma of cinematic temporality itself, film has always been defined by its "crisis" with other media. This persistent anxiety over the essence of the cinematic is not only a fixture of the medium's history, but also a critical component in its larger function, as it is precisely this constitutive crisis that grants the medium its self-critical function. As a result, the medium is uniquely intertwined in ontological fashion with the Bazinian question "What is cinema?," so much so that the identity of this medium, claims Rodowick, is not defined as prior, but immanent, to the audiovisual assemblages that it produces.[17] In the case of *Good Kill*, the convergence between this "ontological groundlessness" of cinema and the media ecology of the kill chain pushes the film's engagement beyond the representational or symbolic layer toward processes of worlding. In this way, the film not only divulges the temporal and spatial convergences that make the global counterinsurgency possible; it also dramatizes the porous boundaries that allow the cinema to retain its currency. From this perspective, *Good Kill* confirms the conflicts that Stewart observes in contemporary film. However, it questions the Manichean relation that such assessments of postcinema often establish between film and contemporary media. With this in mind, the film's apparent lack of a coherent statement appears not so much as a shortcoming, but rather a symptom of its larger mirroring function, which refuses the outside upon which conventional critiques of drone power rely in order to assimilate the kill chain within its processes of worlding.

GROUNDED: THE STAGE AS DIFFERENTIAL MEDIUM

The screen isn't that big
But it becomes your world.
—The Pilot

It is not so much that digital networks destabilize the medium of film, but rather that their horizontal nature helps to reveal its status as always already intermedial. From this perspective, digitality simply dramatizes film's "ontological groundlessness," its articulation of a dynamic and open category that is contingent upon the media ecologies in which it is immersed. And yet it is undeniable that not only are these constitutive modes of convergence often experienced as intrusions into the cinematic experience, but that the larger cultural identity of film also frequently appears at odds with such platforms. This tension can be traced back to the first half of the twentieth century, when techniques of continuity editing combined with discourses of cinematic purity to overtake the off-screen. By way of these techniques, the filmic frame came to delimit an outside that was largely determined by the contents of the image proper. As these practices served to overtake and exclude the noncinematic register from the filmic experience, they reaffirmed a Greenbergian discourse of aesthetic purity in self-fulfilling fashion. From Rudolph Arnheim's rejection of sound to Béla Balázs's dismissal of "photographed theatre," modernism understood film as an autonomous and separate aesthetic category, and this discursive and technological union would ensure this was the case.[18]

While modernistic aesthetics may have fallen out of fashion, residual notions of purity continue to haunt the medium. As a result, the interjection of the space of the theater into the film experience in the 1970s could only be staged as an unraveling rather than a reaffirming of the filmic illusion (importantly, the title of Barthes's seminal essay is "Upon *Leaving* the Moving Theater"). For similar reasons, the migration of the image across digital platforms is rarely presented in terms of a continuation of the medium's proliferation of partial objects and is more often regarded as the corruption of a discrete essence.[19] The effects of this inheritance are not limited to theoretical interventions but are in fact omnipresent in the broader discourse of twenty-first-century cinema, in which the online mash-ups of feature films such as *Star Wars* regularly prompt lawsuits from Hollywood studios, and the widespread closings of neighborhood theaters are often attributed to the rise of digital culture. It is no wonder that one continues to feel a certain interruption or discord when text messages, webcams, or the spatial montages of multiple screens appear in a feature film. While its history and identity is inconceivable without the repetition of this familiar trauma, the cinema nonetheless seems to understand digital media as an unassimilable event.

Theater, on the other hand, achieves its reality through a series of enfoldings by which actors and agencies strategically migrate across the proscenium arch. Most immediately, the soliloquy and the direct address allow characters in mimetic theater to step outside of the diegetic present without necessarily breaking the "fourth wall." However, even the non-Aristotelian theater of Bertolt Brecht or Augusto Boal draws upon these internalized stagings of difference. In fact, Brecht's use of *verfremdungs* to shatter bourgeois ideals implicitly acknowledged the constitutive power of the theater's porosity even as it tore down enduring relations of immersion and identification. In like terms, Boal's "invisible theater" is nothing if not the realization of the theater's infinite flexibility, which allows any time, location, person, or event to comprise a theatrical production. Despite serving different ends and taking vastly different forms, these iterations of the stage articulate their diegesis through a series of convergences by which the medium openly acknowledges a constitutive outside.

This permeability of the onstage diegesis is at the center of the symbiotic relation it maintains to other media. While the Internet is often touted as *the* "universal medium," Chiel Kattenbelt argues that it is the theater that is the true "hypermedium," as it alone maintains the ability to not simply absorb other media, but to do so without reinscribing their mode of representation. Unlike digital networks, which offer a multimedia convergence of sign systems, the theater allows for a physical and material copresence of media. For this reason, it is "the only medium that can incorporate all other media without damaging the specificity of these media and its own specificity."[20] With this assertion, Kattenbelt overturns a familiar point of differentiation between the theater and other media, the liveness of the performer. He argues that the actor's onstage presence is itself inflected by this differential essence such that the performer is simply "the player of the different media who acts in the empty spaces between the media" rather than the authentic presence around which artifice is constructed.[21] Drawing upon this interstitial status of the performer and overall distributed nature of the stage, George Brant's *Grounded* (2013) situates the symmetries of *Good Kill* within a more sustained and coherent critique of drone warfare.

The play tells the story of a female F-16 pilot who is reassigned to Creech Air Force Base to fly drones after returning from maternity leave. Upon receiving her orders, she reluctantly moves to Las Vegas with her child's father, Eric, and the two settle into a typical suburban life. Like her filmic counterpart, "The Pilot" is nagged by a longing to return to flight and soon begins to question the screen-based conflict she is embroiled in. While the incongruity between her day job and civilian life reproduces the conflicts featured in Niccol's film, the power of drone simulation begins to bleed into The Pilot's personal life in more overt and destructive fashion relative to *Good Kill.* And yet these locations, both the base and the domestic sphere, never fully materialize on the stage. In fact, the play consists of an extended monologue delivered by a lone performer who is surrounded by an undulating field of light (see figure 24). While the stage design varies somewhat between productions, the script's declaration that the set "should be more abstract than literal, or perhaps transform the literal into the abstract" persists across each iteration.[22] In this way, "the blue," which serves as a kind of metonymic stand-in for the experience of flight throughout The Pilot's monologue, extends beyond simple mimetic relations with the sky and instead composes a field of possibility from which virtual presences emanate as the performer recounts her experiences as a drone pilot. Its active transformation of the "literal into the abstract" is the condition of possibility for not only the processes of memory, but also the integration of the relations of network-centric warfare into the diegesis of the production.

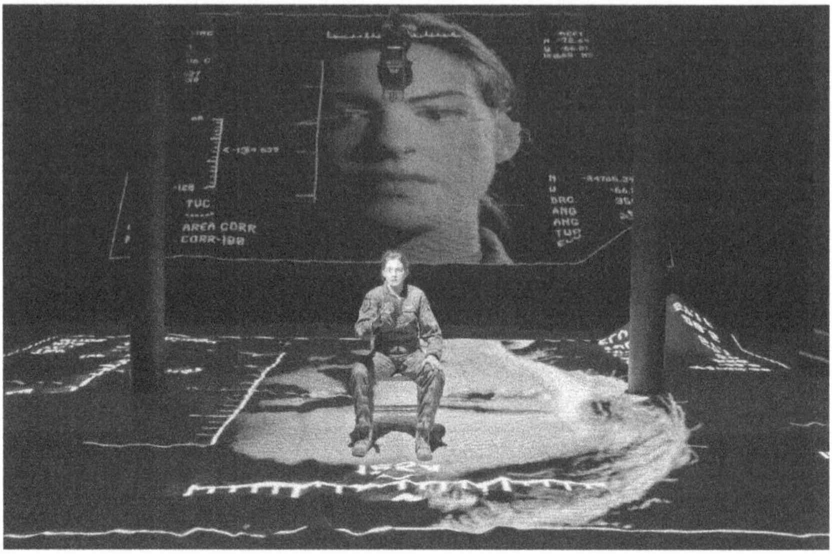

FIGURE 24. Anne Hathaway in *Grounded* at the Public Theater, New York. Photograph: Joan Marcus. © Joan Marcus.

Overseeing these relations are the production's distinct modes of enclosure, which delimit and expand the boundaries of the stage across iterations of the play. These span from the literal, such as in Lisa Portes's American Blues Theater production, which surrounds the performer with monochromatic, unarticulated walls, to the figurative, such as in Wendy Knox's staging for the Playwright Center, which uses light to encase the actress in an implied space that creates an intimate and yet depthless sphere. While these productions interject varying degrees of porosity and enframing, they unanimously conjure an outside that extends beyond physical presence and the mimetic impulse of theater. By eschewing visual devices of reenactment and/or the place-making of traditional set design, this approach allows the narrative relayed by the performer's monologue to achieve a kind of partial presence within these spaces, one that teeters between actuality and virtuality. The effect is that the symbolic representational register of the production, represented by the actor's presence and dialogue, slides into what Francois Lyotard describes as "a general desemiotics of the theater" in which energies, intensities, and libidinal displacements take precedence.[23] Consequently, the stage comes to mark a series of contractions and dilations, articulating simultaneous relations of near and far, past and present. In this way, the play summons not only the contradictory spatial relations of the drone cockpit, but also the powers of the virtual upon which this mode of power relies.

Heiner Müller once described the theater as disclosing "the rebellion of the body against ideas, or more precisely, the impact of ideas and the idea of history upon bodies."[24] The unique physicality that *Grounded* requires of its performer would seem to confirm this relation. While uncoupling bodily presence and physical agency from diegetic space by way of the virtual relations of drone warfare, the theatrical nature of the play at the same time insists that the flesh serve as a primary medium for dramatizing this disconnection. It is on this basis that the nostalgia for flight in *Grounded* takes on an added dimension relative to its filmic counterpart, *Good Kill*. Like Egan, The Pilot's longing for "the blue" is wrapped up with a desire for freedom, sensation, and the kind of expansive landscape that contrasts starkly with the confined stage in which she now resides. And yet "the blue" also operates at a meta-textual level, as it represents the possible reappearance of the sensory motor schema that the virtuality of the drone cockpit (and in turn the diegesis of the play itself) has nullified. One sees this most acutely in Gwendolyn Whiteside's performance. As the following dialogue is recited, she takes on exaggerated bodily gestures, swaying back and forth while tightly pursing her lips, as if muscle memory is the closest she can get to the sensations that she describes. Despite conjuring these mnemonic residues of embodiment, the stage/cockpit will not yield to her desire for an agency that is immediate and nonvirtual. As these nervous ticks essentially perform the antitheatrical nature of drone operation, The Pilot pines:

> It's the speed
> It's the G-Force pressing you back as you tear through the sky
> It's the ride
> My Tiger
> My gal who cradles me lifts me up
> . . .
> You are the blue
> You are alone in the vastness and you are the blue

Following Suzanne Jaeger's definition of the stage as "an active configuring and reconfiguring of one's intentional grasp in response to an environment," one might say that *Grounded* dispenses with the "stage" as a coherent and materially present sphere of action.[25] In its place is a fetishizing of movement so strong that even the recounting of The Pilot's experience of walking with her daughter through the mall becomes an extraordinary event, if for nothing else than the possibility of virtual ambulation ("It's too hot to walk around outside but I need to walk to / move I don't move enough anymore"). These actions not only fail to materialize on stage but in fact appear antithetical to the stark spaces that The Pilot occupies. However, movement and the modes of embodiment it entails do not simply disappear as a result of these relations. Rather, as the performer relinquishes the "intentional grasp" of the stage, the virtual relations of the drone's network come to the surface. As a result, the vague outlines and fleeting impressions that are generated by the inability of bodily action to create worlds create the conditions by which "shape without form" materializes onstage.[26]

In this relationship, screens, which are obsessively referenced in the monologue and insinuated by the ethereal glow of the set, take on a compensatory role. They form "speech objects" that push up against materiality in their claims to presence, both mimicking and unraveling the visible. As these "screens" (always in the plural) hold together the otherwise loose threads of the actress's dialogue, they serve as touchstones, markers by which action is situated: "I watch another screen and I go to bed. . . . [W]e watch the screen call it a night. . . . We stare at a screen and then go to bed." And yet the lack of specificity of these "screens" (Are they TV screens? Tablet screens? Computer screens? The screens of entertainment or the interfaces of war?) erodes the geography of place and the identities of those who would seek to organize it. As their familiar glow illuminates the stage, it subjects the dramatic "action" to the same instability, abandoning the unity of time and place upon which drama is conventionally built.

If screens provide a frame through which discrete agencies collapse in on themselves, then the desert functions as repository and mirror for this process. Proclaiming "out of the suburbs into the desert / The pyramid is behind me," The Pilot repeatedly posits this location as a kind of arbitrator between fields, a threshold that oversees the relation between the simulacra of The Strip and the reality of the suburbs. In this role, its nondescript features provide the raw material for the creation of heterotopias, those layered and stacked dualities that manifest the disruptive symmetries that circulate throughout the play.[27] This status is reaffirmed by interviews with drone pilots, who described the role of the desert in their training in precisely these terms: "Do you understand why they operate in Vegas? Why the military is there: Because the terrain around Vegas, is very similar, and so it is perfect to train drone pilots. When they train one day, and do actual missions the next, they look pretty similar. . . . To practice following a Taliban truck, you practice on tourists going from Los Angeles to Las Vegas on the freeway. Of course, nobody can see them doing that."[28] While *Good Kill* integrates this notion of the Arizona desert as a kind of "mock Afghanistan" into the film through a series of spatial conflations, these interventions are largely confined to the formal dimensions of the work. As a result, the characters themselves are unable to consciously acknowledge the symmetries produced by drone operation even while these relations are readily available to the viewer. *Grounded* similarly utilizes this dual status of the desert to undermine the singularity of the stage and its claims to presence. However, because of its reliance on monologue and abandonment of mimetic sets, the uncertainty regarding the boundaries of the stage necessarily bleeds into the content of the play itself in more overt fashion.[29] For example, while recounting her recent trip to the mall on a day off, The Pilot experiences a sudden breakdown in geographic continuity upon entering the dressing room:

> I go into a changing room
> Take off my shirt
> Look for the camera
> Can't find it
> But there's always a camera right
> JC Penney or Afghanistan
> Everything is Witnessed

Not long afterward, she is assigned to follow the "number two" in command as he rides in a car across the desert. After hours of watching him on the screen, she begins to see herself driving in the arid Nevada landscape. The next day, she returns to the cockpit to follow "her car" on the screen: "Back to Number Two / He just drives / Drives around in my car." This shapeshifting comes to a head at the end of the play, when a young girl runs to the car ("my car"). The target inside is clearly aware of the presence of the drone above and refuses to get out of the car, so as to prevent his identity from being confirmed. As the child runs to the car, "number two" shoos her away to protect her from the possibility of a strike. Suddenly The Pilot sees her daughter on the screen. The voice in her headset commands her to fire, but she is unable to do so. She freezes. A missile is fired, and an explosion engulfs the target and the young girl. Another drone has been hovering above them the whole time.

In the final moments of the play, the ontological slippage that has animated this performance is incorporated into the persona of The Pilot, who no longer speaks as such as she addresses the audience. She declares:

> You who have slaughtered my child
> . . .
> You who have sealed me in this tomb and think you are safe
> Know this
> Know that you are Not Safe
> Know That You Can Keep Me Here Forever You Can Bury Me
> In a Bunker of Grey But That Does Not Protect You For One
> Day it Will Be Your Turn Your Child's Turn and Yea Though
> You May Mark Each and Every Door with Blood None of the
> Guilty Will Be Spared
> None
>
> None
>
> None
>
> boom.

This last word, an onomatopoeia that shatters the veil of virtuality by enacting violence upon the symbolic register of language, lingers as a generic devastation, a disaster to come. One is reminded of Julia Kristeva's description of Celine's representation of a bombing in the novel *Rigadoon*, in which the transcriptions of sound ("boom," "rat-tat-tat," and so on) "'score' and 'punctuate' the utterance, tear it apart."[30] In the case of *Grounded*, this violence enacted upon language serves to recode this monologue with a futural dimension so that speech is divorced from recounting or representing an event and instead becomes the means for dramatizing a chilling threat. In the process, the groundlessness of the virtual infiltrates the performer, causing the specificity of pronouns, events, and identities to break down.

As The Pilot gradually internalizes the ontological disruption of the drone, her performance comes to illustrate not only the capacity of drone strikes to produce terrorists, but also the perpetual nature of a war without boundaries or end. In this regard, this concluding scene drives home the larger conceit of the play to disconnect "radicalization" from religious or ethnic identity and tie it instead to the military actions of the United States. As a result of this reframing, not only do the ethical and legal shortcomings of the global counterinsurgency come into sharp focus, but the fallacy of "bombing our way to safety" also begins to unravel. The relations of postmediality that envelope the diegesis of *Grounded* are integral to this larger statement of the work. In compromising the directionality of conflict, they reveal drone warfare's self-sustaining contradiction, namely, that its operations center as much on the strategic production of insecurity as on the elimination of threats.

CONCLUSION: INTERMEDIALITY BEYOND THE OPERATIONAL

In both *Good Kill* and *Grounded*, the protagonists are subjected to a series of disorienting diegetic folds that confuse the racial, national, and geographic designations that structure drone power. While manifesting as ellipses and overlaps at the level of plot, these folds are in fact the product of a more primary mirroring process by which the multiple media platforms that comprise these theatrical and filmic worlds come to reproduce the relations of the kill chain. The inner workings of this relationship are made apparent when read in conjunction with contemporary discourses of postmediality. Articulating three primary narratives (technical, aesthetic and political), this discourse argues that: as the digital world has eroded medium specificity, it has created a universal medium through which all media flow as nondifferentiated artifacts (Weibel); contemporary intermediality is a symptom of a larger break with modernism's "aesthetic purity," which insisted upon an adherence to the essential properties of a given medium (Krauss); and the universality with which mass media once functioned has given way to distributed ecological formations that preclude singularity (Guattari).[31] The ability of *Good Kill* and *Grounded* to absorb the kill chain into their respective modes of worlding is contingent upon the consolidation of these trajectories. This possibility is not the result of an essential nature of either film or theater, but is rather the product of the precise discourses (the rejection of modernism and the correlative breakdown of medium specificity) and media ecologies (the arrival of distributed media networks) that inscribe these platforms at the historical moment of modern drones.

The relations of sensing that hold together this network of mixed semiotics redistribute the viewer's investment in these works, engendering a mode of spectatorship that is not unlike the experience of apprehending the drone itself. As discussed throughout this book, drones do not present themselves as prosthetic extensions of the audience/operator, but are more immediately legible as external objects. Consequently, they do not ask their users to identify with or "see as" in the same way that the dominant machines of the twentieth century did. Adam Rothstein explains this relationship in the following terms: "For years, the best way to drive a vehicle was with an egocentric perspective—as if the car or plane was us. Similarly, we think about computers and industrial equipment as something that is outside of us, which we look into, control with an interface, and guide around us. But drones change these traditions. Sometimes drone control is easier if we think about it as something we look at, not something we are inside of."[32] This lure to look *at* rather than *through* the drone is succinctly embodied by its blacked out cockpit, which simultaneously affirms both the absence of an onboard operator and the presence of a remote agency. In this way, the visual presence of the drone reiterates the remove that structures drone warfare while at the same time enacting an exceptional closeness for the operator/audience. This simultaneity of near and far, external and internal congeals in the interfaces of drone operation, allowing pilots to shift, most immediately, between egocentric (first person) and exocentric (third person) vantage points, but more broadly between the solidity of presence and the modulated power of absence.

As the drone appears on screen or stage, this pervasive movement serves to interpenetrate form and content in fractal-like fashion, telescoping one through the other such that it is the film or play itself that becomes the subject of the kill chain as much as the reverse. The ensuing collapse of referents and the endless distribution of becoming that runs through these worlds transform the role of the viewer. With identity and geographic location subjected to the permutations of the drone's processes of world-making, one's relation to the screen and/or stage is now defined in terms of a pervasive possibility of violence. As such, the spectators of these works do not so much enter their respective diegeses as experience their own capacity to serve as condition or artifact within martial networks. The sense of apprehension this produces is not unlike the condition of "anticipatory anxiety" that, as chapter 3 describes, is so prevalent among those who live under drones. And yet an attack is not a real possibility for the viewer/citizen of the Global North, and in no way do these works attempt to approximate that experience. In the place of the existential threat of the strike, *Good Kill* and *Grounded* produce bodily manifestations of drone power's strategic animism, the élan vital of network-centric warfare that structures both the material world and its inhabitants. With these processes of becoming target manifesting in the flesh, the body's status as "center of indetermination" becomes the means by which the disruptive power of martial networks is disconnected from the hierarchies of sovereignty and turned back upon itself.[33]

Conclusion

Supersymmetry, Capital, and War

In July 2001, on the eve of 9/11 and the subsequent war on terror, the US Department of Defense delivered a report to Congress that offered a glimpse into the future of the drone program.[1] As the authors of the document reached for a frame of reference that would convey the scale of the transformations in warfare that were about to occur, they turned to a rather curious analogy. Network-centric warfare, the report's authors insisted, is best understood in terms of e-commerce. The document went on to convey the sense of opportunity, the funding structure, and even the risk of this new enterprise in the language of a Silicon Valley start-up pitching to a group of venture capitalists. As the authors pushed this rhetorical strategy further, historical equivalencies emerged that betrayed a larger convergence. At one point, eBay founder Pierre Omidyar was compared to Walter Doran, the navy commander who first implemented network-centric warfare as part of the 1998 Fleet Battle Experiment (FBE).[2] Even the underlying asymmetry of drone warfare was posited as a reiteration of the capitalist enterprise: "The competitors who were first able to correctly identify the opportunity space provided by the Internet and e-business have been able to reap disproportionate rewards. The DoD seeks similar disproportionate advantages in future conflicts as we develop and implement a strategy for a network-centric transformation."[3]

War profiteering is as old as conflict itself. As S. L. A. Marshall, the chief combat historian for the military during World War II and the

Korean War, rather crassly put it, "battle is the freest of all free enterprises."[4] If the exorbitant profits claimed by companies such as Haliburton, HSBC, and Global Risk Strategies are any indication, recent conflicts are no exception. However, the motivations behind the report's analogy with e-commerce extend beyond this age-old drive. Rather than being a simple promise of return on investment, the analogy speaks to a larger vision of drone power in which the martial networks that undergird these operations would be seamlessly integrated with both online civilian platforms and the trajectories of late capitalism that enframe them. The end goal of this convergence is envisioned in the following passage:

> [T]he terms "Network Centric Operations" and "NCW" [network-centric warfare] are used to describe various types of military operations in the same way that the terms "e-business" and "ecommerce" are used to describe a broad class of business activities that are enabled by the Internet. Scott McNealy, chairman and CEO of Sun Microsystems, recently stated, "The "e" in e-business is redundant." His basic point is that e-business has to be about creating value and making a profit or it is not going to be relevant. In a similar sense, NCW is very much about warfare employing Information Age concepts to increase combat power in war and mission effectiveness in operations other than war.[5]

In hindsight, it is clear that the disappearance of drones into mediascapes "other than [that of] war" comprised the ultimate horizon for this technology from its inception. In order to realize this fantasy, an entanglement between digital capitalism and martial networks has run parallel to the war on terror over the past decade and a half. As a result of this shadow narrative, the interpenetration of war and peace, which was isolated to certain geographic areas under initial modes of drone power, is increasingly becoming a shared global everyday. While drone art's relation to this new modality of martial power remains tentative and unresolved, the genre has nonetheless begun to register the challenges this prospect introduces for its accelerationist program.

Nick Srnicek and Alex Williams begin their "#ACCELERATE MANIFESTO for an Accelerationist Politics" by disentangling the affirmationist tendencies of accelerationism from the techno-utopian leanings of twentieth-century media activism. They do this by problematizing the notion that technology and the social field operate independently of one another. Rather than a progressive social or political agenda utilizing technology for its realization, accelerationism, they insist, understands the social as always already technological.[6] While not necessarily isolated to the contemporary, this condition is nonetheless amplified by the expansion of the scope and scale of subsumption that occurs under the digital economy of late capitalism. Shifting from what Shaviro describes as extrinsic to intrinsic exploitation, the expenditures of labor are now directly incorporated as commodity via social media and the modulated surveillance of big data. As a result, "production and circulation [have become] indistinguishable" and the mode of communication serves as its own content.[7] These relations materialize as "communicative capitalism," a fantasy fueled by techno-fetishism and illusory models of participation, which together present digital culture as the realization of democratic ideals of an engaged citizenry.[8]

The incestuous relationship between civilian media and warfare that the DoD envisioned in the weeks before 9/11 has increasingly become reality by way of this system. Not only does communicative capitalism's emphasis on exchange bolster intelligence gathering, but so do the locational data of mobile devices operate as a primary means of targeting, while social media platforms provide the basis for the internal operations of the kill chain. This overlap is perhaps most pronounced in initiatives such as Project Maven, in which the DoD has recruited Google to bring its systems of machine learning and object recognition to the "millions of hours" of footage accrued by drones.[9] Such Algorithmic Warfare Cross-Functional Teams (AWCFT's) are exemplary of what Manuel Castells describes as "meta-networks," joint formations that amplify the power and reach of individual networks via a combinatory "switching power" while at the same time appearing to preserve unique spheres of action.[10] Through these formations, everyday communication and affective states are not simply commodified but militarized via the global counterinsurgency and its expansive networks. This condition forms the basis of what James der Derian calls the "military-industrial-media-entertainment network" (MIME-NET), which escalates the intrusion of the "living room war" of Vietnam to a war waged within the global networks that constitute daily life.[11]

The sheer diversity of the roles that drones play in contemporary culture is a testament to this increasing integration into civilian life. UAVs are now used to prevent graffiti in train yards, deter poachers, monitor hurricanes, study crops, track ice floes, execute search and rescue, disarm explosives, assess the safety of mines, prospect for oil, provide emergency medical support, assist in border patrol, and provide intelligence to law enforcement. Perhaps most notable of these nonmilitary markets is the burgeoning consumer drone industry, which sold 2.4

million units in the United States alone in 2016 and posted revenues of over $4 billion.[12] The stakes of this assimilation of drone power into the everyday are perhaps most legible in the seemingly banal practice of the "dronie," a subgenre of the selfie in which users employ the consumer drone to generate panoramic self-portraits. While appearing to echo the mirroring operations of drone art, this mode of symmetry seems utterly devoid of the latter's perversity. The conjoining of the genre of self-portraiture with the visual grammar of surveillance and state violence here represents less a dynamic fold than a congealing mash-up, which effectively reterritorializes the network. It is not simply the capacity of "dronies" to normalize the technology, but the way in which these practices recast the fluid movement between the vertical sovereignty and decentralized networks as a directional relation, that so flippantly places the latter in the service of the former. In this way, the dronie portends a static symmetry on the horizon, a simultaneity based not on clandestine modes of modulated becoming, but rather a particular formulation of transparency through which the weapons of war achieve invisibility by virtue of their indistinguishability from the tools of capitalism. This latter possibility is fueled in part by a dialectical conception of transparency, which Slavoj Žižek attributes to the late-twentieth-century ascendancy of the interface over the machine:

> Modernist technology is "transparent" in the sense of retaining the illusion of an insight into "how the machine works"; that is to say, the screen of the interface was supposed to allow the user direct access to the machine behind the screen. . . . The postmodernist "transparency" designates almost the exact opposite. . . . [T]he interface screen is supposed to conceal the workings of the machine, and to simulate our everyday experience as faithfully as possible. However, the price of this illusion of a continuity with our everyday environs is that the user becomes "accustomed to opaque technology"—the digital machinery "behind the screen" retreats into total impenetrability, even invisibility.[13]

In these terms, drones do not disclose their inner workings as they enter into everyday environments, but rather ensure their own opacity through these new relations of firsthandedness. As a result, this pursuit of "total impenetrability" is not only not opposed to the transparency of civilian media, but is increasingly unimaginable outside of the online everyday of communicative capitalism.

The possibility of this fully assimilated mode of drone power has structured the twenty-first-century evolution of this technology, even while the latter has preserved the Global North's temporary state of exception. As a result of this relationship, state violence articulates both the everywhere war and the elsewhere war, a contradictory status in which drone power uses the global everyday as the basis for its operations while exercising violence in purely asymmetrical terms. As this book has illustrated, the tensions of such a project have provided critical entry points for drone art, which has sought to leverage these relations by presenting not only partial objects that await activation via civilian distributed networks (networked), but also artifacts that have internalized the pliability of digital media into their form (objectile). Conjoining platforms such as 3-D printing, photography, installation, and video, these objects initiate a dynamic web of iterations, forming a chronology of shape-shifting that echoes drone power's abandonment of identity and representation as the anchoring determinants of martial force. These strategies are often bolstered by alternate histories of the UAV, which leverage the nonchronologies that reside within the interfaces and procedures of drone operation (historiographic). As the inherent asymmetries of martial networks serve as the conditions of possibility for their reconfiguration, this first phase of drone art articulates a relation to military technology that is steeped in what Ulises Mejias calls "parasitology." He describes this relationship in the following terms:

> To unthink the logic of the digital network is not to refuse to confront the network, pretending it does not exist, but to reimagine one's relationship to it. The relationship of the outside to the inside might then be like that of the parasite to the host, if we consider those arguments about how the parasite inserts itself into the communication network between two nodes—the sender and the receiver—disrupting the flow of information by adding noise (information outside the logic of the system) and forcing the network to adjust to its presence.[14]

In like terms, drone art's assimilation of civilian networks is staged less as an intervention than as an escalation of latent formations already operating within the kill chain and its ecology of agencies. By introducing new configurations via secondary nodes or dead spots, those spaces or intervals where forgotten technologies and unrealized social and political formations reside, this work attempts to invert the system's relation to noise so that "communication in spite of noise is replaced with communication through noise."[15] Championing "counter-protocological" practices and an overall "informatics" of thinking, these practices echo what Galloway and Thacker describe as the utilization of nonhuman actors to establish "super-pliant" networks within networks that "exploit" the power differentials of the sovereignty-network dialectic. These postideological interventions produce a point of entry for what the authors call "the exploit," a strategy that utilizes the asymmetry of the network to push it toward inverse manifestations.[16] As a genre, drone art has up until this point internalized an oscillation or movement between these formations, both redirecting from within by accelerating power differentials to the point of deformation and at the same time imagining sustainable modes of communication from these same relations. Through this movement, potentialities gather and materialize along a spectrum of overlapping, contradictory networks, producing a deterritorializing force that operates alongside and within modulated fields of power.

However, in the face of the kind of absolute assimilation envisioned by contemporary drone power, such leverage points may prove increasingly unavailable. After all, the parasite/host relationship relies on a hierarchy of nodal centers that may no longer be legible in this environment. The shifting valence that strategies of perverse symmetry assume in this context are dramatized by the following remarks by artist Suzanna Treister.

> [T]aking surveillance culture as a given and looking at the world where it is a given and thinking from there, I can try to present what this culture looks like, its imagery, its vocabulary, its aspirations and its potential futures, to reflect a post-surveillance mindset. The work is ironically part accepting and part moving on, and sometimes celebratory. What if we stop complaining/justifying for a minute, so boring, let's try and visualise this brave new world. The works also retain an inbuilt criticality; you could say they are ambiguous in their political position. I like that. I like the idea of playing with an image of the NSA, or GCHQ, who store all our data, and representing it all as a kind of hallucinogenic drug induced visionary landscape, a kind of pop poetry. I like imagining the edge of the future.[17]

Without explicitly referencing accelerationism, the artist's privileging of "reshaping and repurposing (rather than refusing)" and overall sense of deferral to processes of martial power betray an affinity with this aesthetic strategy.[18] However, what is noteworthy in these remarks is the extraordinary ambiguity they ascribe to these operations. Admitting to a conflicted attitude toward her source material and the larger possibility of resistance, the artist seeks out a curious disposition in which "accepting" would no longer be opposed to reimagining. As the din of activists and apologists seems equally inadequate to this "post-surveillance mindset" ("What if we stop complaining/justifying?"), she insists that a radical departure is in order, a "moving on" that might somehow extend the "death of the author" to both artist and audience alike. The impending relations of supersymmetry ("surveillance culture [is] a given") would therefore seem to require that the "unmanned" aspect of drones bleed over into the critical operation itself, manifesting as a glaring antihumanism within thought. In this regard, the artist's notion of "inbuilt criticality" proposes a deeper parallelism, one that seeks to expand beyond intentional and human-centered modes of reproduction to conjure "the edge of the future." Treister's installation, *The Drone That Filmed the Opening of Its Own Exhibition* (2013), sheds light on this prospect.

Sitting motionless on top of a white pedestal, the work's Parrot drone is immediately coded as an inert object as one enters the gallery. However, as a video feed appears on a nearby monitor, the viewer soon realizes this object is also the origin of a gaze (see figure 25). This familiar tension between object and medium escalates as the drone rises from the plinth and moves through the gallery. As it does, the paintings and photographs that hang on nearby walls appear to migrate onto the screen, granting the drone a role that exceeds that of a simple guide or docent. It soon becomes clear that this drone does not perceive as the viewer, nor does it act as a prosthetic extension of his or her sensorium. The sensation is rather a perceiving in the place of or even at the expense of the subject. The viewer of the work is thus externalized not by the powers of fantasy and misrecognition, but by the mobility, real-time processing, and overall sentience of a networked presence. While similar relations characterize the works discussed in this book, Treister's piece foregrounds this surrender of agency in more explicit fashion and in so doing replaces the disruptive capacity of symmetry with the relations of "interpassivity" discussed in chapter 3. This outsourcing of the gallery experience to the drone serves not only to dramatize the increasing collapse of communicative capitalism with martial power, but also to call into question the disruptive potential of accelerationist tactics of symmetry in the face of this emerging system.

As the drone acknowledges visitors in the gallery, the wall-mounted screen displays exceptionally self-aware subjects. Shunning true interactivity in favor of a compulsory visibility, the work effectively peels off the "data double" from the perceiving self, prompting a free flow of signifiers.[19] The exhibitionist performances that ensue are oddly familiar in their banality and anticlimax. In fact, one is struck by the way in which this flow of images reproduces the stock subjectivities of selfies, memes, and cell phone videos that populate social media. In generating

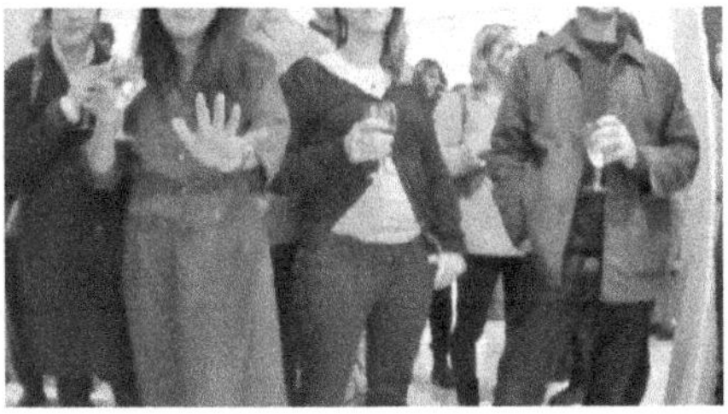

FIGURE 25. Suzanne Treister, *The Drone That Filmed the Opening of Its Own Exhibition*, 2013. Video still. Courtesy of the artist.

these artifacts, the drone no longer functions as a passive surveillance apparatus that looks in from the outside, but rather becomes indecipherable from the medium it engages. This conflation echoes recent military practice of equipping drones with "virtual base-tower transceivers" that act as "false cellphone towers" as they enter a given airspace. Rerouting data traffic in the area through its servers in order to harvest valuable intelligence, the drone serves in this instance as both the foundation for everyday digital media and, as Lisa Parks puts it, the "digital vacuum cleaner" of the kill chain.[20] Through this entanglement of the externalized subjectivity of communicative capitalism with the operations of martial networks, the work materializes the impasse that supersymmetry poses. The impenetrability of this mode of power culminates in the drone itself, which appears decidedly nonconfrontational and domesticated in Treister's work. Having absorbed the asymmetries of its earlier iterations, this object is no more foreign or disruptive than a doorbell that broadcasts a live video feed to one's phone or a voice-activated speaker that responds to its name. Whatever semiotic charge it retains seems to emanate from its status as consumer electronic and the unique lure of interpassivity that such devices articulate rather than from its history as a weapon of war.

The first generation of drone art addressed the question of what happens when a drone enters the gallery, appears on the screen, or populates urban spaces. Leveraging the tensions provoked by such scenarios, this work sought to preserve these questions as such in order to manifest and exploit unsettling relations of symmetry. Now, as relations of supersymmetry manifest between martial networks and civilian media, a reconsideration of these tactics is in order. As these networks of violence progress toward indistinguishability within civilian platforms, it appears likely that the collision of drone and social media mobilized by the work of Bridle, Paglen, and others will not retain the same capacity for disruption. Indeed, as we begin to control our own drones with smartphones, the targeting of subjects via SIM cards will appear slightly less outrageous. The presence of drone silhouettes in online satellite maps or even galleries and museums will no longer seem so out of place. At the same time that these processes of normalization take hold, the skies above the Global North will begin to fill with drones. As global powers lose their monopoly on unmanned airspace and drones from other countries, corporations, or "non-state actors" begin populating the atmosphere, how will those below disentangle the contradictions of a medium that is equipped to deliver both packages and bombs? Are there alternate symmetries that might be called upon to distort these relations into unrecognizable perversions of themselves? Will new modes of interpassivity take on the disruptive capacity of their predecessors, or will this delegation to the machinic only serve to mourn the disappearance of an outside to drone power?

Notes

INTRODUCTION

1. Holland Cotter, "'Laura Poitras: Astro Noise' Examines Surveillance and the New Normal," *New York Times*, February 4, 2016.

2. The editors of *October* go on to suggest that these conditions have reduced "the role of academics, intellectuals, and artists in the cultural public sphere . . . to anesthesia and amnesia." Cited in Rosalyn Deutsche, *Hiroshima after Iraq: Three Studies in Art and War* (New York: Columbia University Press, 2010), 2.

3. David Hambling, "Game Controllers Driving Drones, Nukes," *Wired*, July 17, 2008, www.wired.com/2008/07/wargames/.

4. For more information on this project, see Thomas Stubblefield, "'Non-Offensive Computing: Project Maven and Google's Discourse of Drone Power," in *Media Reflections: Threat Discourse and the American Order Since the Attacks of September 11, 2001* (New York: Routledge, forthcoming).

5. The term *mediality* is used to acknowledge a shift from regarding media as autonomous means of communication toward a more expansive view in which they are understood in terms of a dynamic set of relations that actively comprise the conditions that make communications and their respective realities possible. Relative to "media," mediality represents a more active category that encompasses the connections of a given medium to other media as well as the myriad of processes of mediation that take place beyond the confines of a given "text." See Jørgen Bruhn, "What Is Mediality, and (How) Does It Matter? Theoretical Terms and Methodology," in *The Intermediality of Narrative Literature: Medialities Matter*, ed. Jørgen Bruhn (New York: Palgrave, 2016), 13–40.

6. Alexander Galloway and Eugene Thacker, *The Exploit: A Theory of Networks* (Minneapolis: University of Minnesota Press, 2007).

7. Jason Bittel, "Want a Drone-Hunting License in Colorado? Get in Line," *Slate*, September 6, 2013, www.slate.com/blogs/future_tense/2013/09/06/deer

_trail_co_sees_applications_for_nonexistent_drone_hunting_license.html. See also Greg Campbell, "Drone on the Range: A Colorado Town Goes Hunting for Unmanned Aircraft," *Al Jazeera*, December 15, 2013, http://america.aljazeera.com/articles/2013/12/15/drones-on-the-rangeacoloradotowngoeshunting.html.

8. Mark Fisher succinctly summarizes the structure of this disavowal in the following terms: "So long as we believe in our hearts that capitalism is bad, we are free to participate in capitalist exchange." *Capitalist Realism: Is There No Alternative?* (London: Zero Books, 2009), 13.

9. See Lauren Berlant, *Cruel Optimism* (Chicago: University of Chicago Press, 2011).

10. The phrase "everywhere war" refers to a borderless mode of conflict that understands the "enemy" as a floating designation that can be extended to almost any subject regardless of location or national affiliation. Derek Gregory, "The Everywhere War," *Geographic Imaginations* 177, no. 3 (2011): 238–250.

11. Michael Dillon and Julian Reid, *The Liberal Way of War: Killing to Make Life Live* (New York: Routledge, 2009).

12. Cited in Samuel Weber, "Target of Opportunity: Networks, Netwar, and Narratives," *Grey Room* 15 (Spring 2014): 7.

13. Cited in Ulises Ali Mejias, *Off the Network: Disrupting the Digital World* (Minneapolis: University of Minnesota Press, 2013), 92.

14. Alex Williams, "Escape Velocities," *E-Flux*, no. 46 (June 2003), www.e-flux.com/journal/escape-velocities/.

15. Steven Shaviro, "Accelerationist Aesthetics: Necessary Inefficiency in Times of Real Subsumption," *E-Flux*, no. 46 (June 2013), www.e-flux.com/journal/46/60070/accelerationist-aesthetics-necessary-inefficiency-in-times-of-real-subsumption/.

16. Robin Mackay and Armen Avanessian, *The Accelerationist Reader* (London: Urbanomic Press, 2014), 23.

17. Dane Sutherland, "Vaporents: Inhuman Orientations," *Culture Machine* 16 (2015): 16.

18. See part II of Mejias, *Off the Network*, 80–144.

19. Gilles Deleuze, *The Fold: Leibniz and the Baroque* (London: The Athlone Press,1993), 20.

20. Graham Harwood, "Endless War: On the Database Structure of Armed Conflict," *Rhizome*, March 17, 2014.

21. Karen Beckman, "Telescopes, Transparency, and Torture: Trevor Paglen and the Politics of Exposure," *Art Journal* 66, no. 3 (2014): 62.

22. Arthur K. Cebrowski and John J. Garstka, "Network-Centric Warfare: Its Origin and Future," *Proceedings of the U.S. Naval Institute*, January 1998, www.usni.org/magazines/proceedings/1998/january/network-centric-warfare-its-origin-and-future.

23. Cited in Deutsche, *Hiroshima After Iraq*, 2.

24. Caren Kaplan, "Everything Is Connected: Aerial Perspectives, the Revolution in Military Affairs, and Digital Culture," *HASTAC*, August 18, 2009, www.hastac.org/electronic-techtonics/caren-kaplan-everything-connected-aerial-perspectives-revolution-military.

25. Mike Hill, "Ecologies of War: Dispatch from the Aerial Empire," in *Telemorphosis: Theory in the Era of Climate Change*, Vol. 1, ed. Tom Cohen (Ann Arbor, MI: Open Humanities Press, 2012), https://quod.lib.umich.edu/o/ohp/10539563.0001.001/1:13/--telemorphosis-theory-in-the-era-of-climate-change-vol-1?rgn=div1;view=fulltext.

26. Conor Friedersdorf "Every Person Is Afraid of the Drones: The Strikes' Effect on Life in Pakistan," *Atlantic,* September 25, 2012, www.theatlantic.com/international/archive/2012/09/every-person-is-afraid-of-the-drones-the-strikes-effect-on-life-in-pakistan/262814/.

27. Rosalind Krauss, *A Voyage on the North Sea: Art in the Age of the Post-Medium Condition* (London: Thames & Hudson, 1999), 64.

28. Quoted in Patrice Pavis, "The State of Current Theatre Research," *Applied Semiotics/Sémiotique Appliquée* 1, no. 3 (1997): 203.

29. William Gibson, *Spook Country* (New York: Penguin, 2007), 63.

1. SIGNATURE STRIKES AND WORLD-MAKING

1. Sharon Weinberger, "How ESPN Taught the Pentagon to Handle a Deluge of Drone Data," *Popular Mechanics*, June 11, 2012, www.popularmechanics.com/technology/military/planes-uavs/how-the-pentagon-will-handle-its-deluge-of-drone-data-9600910.

2. Cora Currier and Justin Elliot, "The Drone War Doctrine We Still Know Nothing About," *ProPublica*, February 26, 2013, www.propublica.org/article/drone-war-doctrine-we-know-nothing-about. See also Greg Miller, "Proposal to Give Federal Judges a Role in Drone Strikes Faces Hurdles," *Washington Post*, February 8, 2013, www.washingtonpost.com/world/national-security/proposal-to-give-federal-judges-a-role-in-drone-strikes-faces-hurdles/2013/02/08/66f53508-721a-11e2-8b8d-e0b59a1b8e2a_story.html.

3. Analyzing the organizational structure of the 9/11 terrorists, Valdis Krebs identifies three central components of such networks: incompleteness, fuzzy boundaries, and dynamic relations. Valdis Krebs, "Mapping Networks of Terrorist Cells," *Connections* 24, no. 3 (2002): 43–52.

4. Steve Ressler, "Social Network Analysis as an Approach to Combat Terrorism: Past, Present, and Future Research," *Homeland Security Affairs* 2, article 8 (July 2006): 2.

5. Mark Poster, *The Mode of Information: Poststructuralism and Social Context* (Chicago: University of Chicago, 1990), 97–98.

6. "US Drone Strikes Kill 28 Unknown People for Every Intended Target, New Reprieve Report Reveals," Reprieve.org, November 25, 2014, https://reprieve.org.uk/press/2014_11_25_us_drone_strikes_kill_28_each_target/. See also Reprieve, "You Never Die Twice: Multiple Kills in the US Drone Program," November 24, 2014, www.reprieve.org/wp-content/uploads/2014_11_24_PUB-You-Never-Die-Twice-Multiple-Kills-in-the-US-Drone-Program-1.pdf.

7. These intertwined dynamics are confirmed by Matthew Power's article on drone pilots for *GQ*: "He was told that they were carrying rifles on their shoulders, but for all he knew, they were shepherd's staffs. Still, the directive from somewhere above, a mysterious chain of command that led straight to his headset,

was clear: confirmed weapons. . . . Bryant's laser hovered on the corner of the building. 'Missile off the rail.' Nothing moved inside the compound but the eerily glowing cows and goats. Bryant zoned out at the pils. Then, about six seconds before impact, he saw a hurried movement in the compound. 'This figure runs around the corner, the outside, toward the front of the building. And it looked like a little kid to me. Like a little human person.' Bryant stared at the screen, frozen. 'There's this giant flash, and all of a sudden there's no person there.' He looked over at the pilot and asked, 'Did that look like a child to you?' They typed a chat message to their screener, an intelligence observer who was watching the shot from 'somewhere in the world'—maybe Bagram, maybe the Pentagon, Bryant had no idea—asking if a child had just run directly into the path of their shot. 'And he says, 'Per the review, it's a dog.' Bryant and the pilot replayed the shot, recorded on eight-millimeter tape. They watched it over and over, the figure darting around the corner. Bryant was certain it wasn't a dog." Matthew Power, "Confessions of a Drone Warrior," *GQ*, October 23, 2013, www.gq.com/story/drone-uav-pilot-assassination.

8. Levi R. Bryant, *The Democracy of Objects* (Ann Arbor: University of Michigan/Open Humanities Press, 2011). The term *flat ontology* can also be traced back to Manuel DeLanda: "[W]hile an ontology based on relations between general types and particular instances is hierarchical, each level representing a different ontological category (organism, species, genera), an approach in terms of interacting parts and emergent wholes leads to a flat ontology, one made exclusively of unique, singular individuals, differing in spatio-temporal scale but not in ontological status." *Intensive Science & Virtual Philosophy* (London: Bloomsbury Press, 2013), 47.

9. Mike Hill, "Ecologies of War: Dispatch from the Aerial Empire," in *Telemorphosis: Theory in the Era of Climate Change*, vol. 1, ed. Tom Cohen (Ann Arbor, MI: Open Humanities Press, 2012), https://quod.lib.umich.edu/o/ohp/10539563.0001.001/1:13/--telemorphosis-theory-in-the-era-of-climate-change-vol-1?rgn=div1;view=fulltext.

10. While not intended as a direct correlate, Nelson Goodman's use of the term *world-making* and his larger theory of the multiplicity of worlds is nonetheless instructive in understanding both the operation of the kill chain and the interventions that artists have made in these relations. Insisting that there is no universal or stable world out there for us to experience ("content vanishes without form"), Goodman posits the subject as primary to the creation of worlds. Yet this process does not begin from the abyss, that familiar trope of French philosophy, but rather from other worlds ("the search for a universal or necessary beginning is best left to theology"). Accordingly, a primary dynamic in this process of world-making for Goodman is the processes of composition and decomposition. While the ontological leap of signature strikes may extend beyond the purview of Goodman's original formulation, the processes by which worlds are constructed remains relevant. It is on this basis that the twenty-first-century iteration of the drone distinguishes itself from its historical predecessors such as the Firebee, used during the Vietnam War or the Kettering bug of World War I. Nelson Goodman, *Ways of World-Making* (Indianapolis, IN: Hackett Publishing, 1978), 2–5.

11. William James, *Principles of Psychology* (New York: Dover Publications, [1890] 1950), 95–96. The original quote can be found in George John Romanes, *Mental Evolution in Animals: With a Posthumous Essay on Instinct by Charles Darwin* (New York: D. Appleton and Company, 1884), 324. See also Hanna Rose Shell, "The Crucial Moment of Deception," *Cabinet* 33 (Spring 2009), www.cabinetmagazine.org/issues/33/shell.php.

12. Henri Bergson, *Matter and Memory* (New York: Zone Books, 1988).

13. In 2008 alone, the US military accumulated twenty-four years' worth of drone video. However, a massive cache of data leaked by Chelsea Manning and made public by Wikileaks in 2010 suggests that this abundance of visual information may actually be overshadowed by its nonvisual counterpart. Referred to as the *Afghan War Diary*, this archive contains more than ninety thousand documents that span an array of fields, including human intelligence (HUMINT), psychological operations (PSYOP), engagement, counter improvised explosive devices (CIEDs), significant acts (SigActs), targeting, and social and cultural reports.

14. Derek Gregory, "Targeted Killings and Signature Strikes," *Geographical Imaginations*, November 6, 2012, http://geographicalimaginations.com/2012/11/06/targeted-killings-and-signature-strikes/.

15. David S. Cloud, "Anatomy of an Afghan War Tragedy," *LA Times*, April 10, 2011, http://articles.latimes.com/2011/apr/10/world/la-fg-afghanistan-drone-20110410. Such transubstantiation is part of a larger program of strategic shape-shifting that serves not only to necessitate a future attack but also to justify past actions. The latter is confirmed by Airman First Class Brandon Bryant, who describes how a child seen on the monitor moments before a drone strike was dismissed as a dog in subsequent reports of the incident (see note 7). Summarizing the fluid nature of these signifiers, interviewer Matthew Power similarly asserts that the alleged rifles carried by the subjects of drone surveillance could just as easily have been shepherd's staffs and vice versa. Power, "Confessions of a Drone Warrior."

16. Graham Harwood, "Endless War: On the Database Structure of Armed Conflict," *Rhizome*, March 17, 2014, http://rhizome.org/editorial/2014/mar/17/endless-war-database-structure-armed-conflict/.

17. Harwood, "Endless War."

18. As discussed in chapter 3, the widespread belief among Pakistanis that SIM cards emit signals to drones or that informants have deposited "US-provided 'transmitter chips' at local Taliban and Al Qaeda houses [in order to] signal specific targets for [Central Intelligence Agency] CIA drone strikes" testifies to the extraordinary agency that objects attain as a result of these relations, an agency that does not so much implicate an innocent subject through misrepresentation as create a world in which he or she no longer appears as such. The effects of this system are affirmed by the Borgen Project's assessment of the impact of drones on the mental health of Pakistanis: "Doctor Muktar ul-Haq is the head of psychiatry at a government hospital in the city of Peshawar. He told a story of a man who had a full blown psychotic episode after he found a SIM card outside of his house. "Drones Impact on Mental Health in Pakistan," *The Borgen Project* (blog), December 31, 2013, http://borgenproject.org/drones-impact-mental-health-pakistan/.

19. Jordan Crandall, "Un-manned," in *Manifestation & Materialization: UFO Drag Queen; Delirium* (Atlanta, GA: Fort!/Da! Books, 2011), 108.

20. Harwood, "Endless War."

21. Simone Weil, *Gravity and Grace* (New York: Psychology Press, 2002), 22.

22. Samuel Weber, *Targets of Opportunity: On the Militarization of Thinking* (New York: Fordham University Press, 2009), 4.

23. Weber, *Targets of Opportunity*, 5.

24. David Ensor, "Drone May Have Spotted bin Laden in 2000," CNN, March 17, 2004, www.cnn.com/2004/WORLD/asiapcf/03/17/predator.video/.

25. On that day, a prototype for the Predator successfully fired a Hellfire missile at a target at Nellis Air Force base for the first time.

26. Derek Gregory, "The Everywhere War," *Geographic Journal* 177, no. 3 (September 2011): 239.

27. Johanna Drucker, "Humanities Approach to Interface," *Culture Machine* 12 (2011): 9.

28. Victor Stoichita, *The Self-Aware Image: An Insight into Early Modern Meta-Painting* (Cambridge, UK: Cambridge University Press, 2001).

29. Drucker, "Humanities Approach to Interface," 9.

30. Theodor Adorno, *Minima Moralia*, Marxists.org, [1951], www.marxists.org/reference/archive/adorno/1951/mm/ch01.htm.

31. Alex Leveringhaus and Tjerk de Greef characterize the automation of contemporary targeting systems according to four criteria: the ability to "(1) acquir[e] information about an environment via sensors, (2) analyz[e] information, (3) mak[e] a decision about taking a specific course of action, and (4) execut[e] the decision." Within each of these layers, human operators maintain varying degrees of intervention. They can be "in the loop" (remotely controlled), "on the loop" (operator on standby), or "out of the loop" (operator absent). This schema is a productive means not only of analyzing various modes of drone operation, but also of framing Garnett's revisions to the traditional viewing position of realist painting. To put these relations in terms of Leveringhaus and Greef's schema, one might say that the peculiar position of the viewer/operator is no longer that of being "in the loop" but rather "on the loop." However, the real impact of these works is their rejection of the binary in which inorganic agencies "analyz[e] information" and human operators "execut[e] decisions." Instead, the *Predator* series visualizes a deep interpenetration of agencies by way of an enlivened materiality. Alex Leveringhaus and Tjerk de Greef "Keeping the Human 'In the Loop': A Qualified Defense of Autonomous Weapons," in *Precision Strike Warfare and International Intervention: Strategic, Ethico-Legal, Decisional Implications*, ed. Mike Aaronson, Wali Aslam, Tom Dyson, and Regina Rauxloh (New York: Routledge, 2014).

32. Mark Dorrian, "Drone Semiosis: Weaponry and Witnessing," *Cabinet* 54 (Summer 2014).

33. Nathaniel Hartman, "Asm_frag," accessed June 12, 2017, http://nathanielhartman.com/portfolio/asm_frag/.

34. Hartman, "Asm_frag."

35. Hartman, "Asm_frag."

36. Jo Becker and Scott Shane, "Secret 'Kill List' Proves a Test of Obama's Principles and Will," *New York Times*, May 29, 2012, www.nytimes.com/2012/05/29/world/obamas-leadership-in-war-on-al-qaeda.html.

37. The drive of the counterinsurgency to abandon ends-based, linear campaigns in favor of the antinarrative of perpetual war bears all the markings of the database's perpetual present.

38. She explains: "Data are familiarly 'collected,' 'entered,' 'compiled,' 'stored,' 'processed,' 'mined,' and 'interpreted.' Less obvious are the ways in which the final term in this sequence—interpretation—haunts its predecessors. At a certain level the collection and management of data may be said to presuppose interpretation. . . . Data need to be imagined as data to exist and function as such, and the imagination of data entails an interpretive base." Lisa Gitelman and Virginia Jackson, introduction to *"Raw Data" Is an Oxymoron* (Cambridge, MA: MIT Press 2013), 3.

39. Jutta Weber, "Keep Adding: On Kill Lists, Drone Warfare and the Politics of Databases," *Society and Space* 34, no. 1 (2016): 112.

40. "Intelligence Analysis," IBM, accessed July 1, 2016, www.ibm.com/security/intelligence-analysis.

41. Brendan McGarry, "Special Forces, Marines Embrace Palantir Software," *DefenseTech*, July 1, 2013, www.defensetech.org/2013/07/01/special-forces-marines-embrace-palantir-software/.

42. Weber, "Keep Adding," 112.

43. Bernhard Rieder, "ORDER BY column_name: The Relational Database as Pervasive Cultural Form" (presented at The Lived Logics of Database Machinery Conference, London, 2012).

44. Gareth Porter, "How McChrystal and Petraeus Built an Indiscriminate Killing Machine," Truthout, September 26, 2011, www.truth-out.org/news/item/3588:how-mcchrystal-and-petraeus-built-an-indiscriminate-killing-machine.

45. Weber, "Keep Adding."

46. Glenn Greenwald and Jeremy Scahill, "The NSA's Secret Role in the U.S. Assassination Program," *Intercept*, February 10, 2014, https://firstlook.org/theintercept/2014/02/10/the-nsas-secret-role/.

47. Katherine Hayles, *How We Think: Digital Media and Contemporary Technogenesis* (Cambridge, MA: MIT Press, 2012), 177.

48. Hayles, *How We Think*, 177–178.

49. Porter, "How McChrystal and Petraeus Built an Indiscriminate Killing Machine."

50. Joseph Pugliese, "Drone Casino Mimesis: Telewarfare and Civil Militarization," *Journal of Sociology* 52, no. 3 (2016): 503.

51. It is clear that this system makes full use of the fact that, as Schmid describes, there is no universally recognized or legally binding definition of terrorism, and in fact the definitions vary dramatically between agencies even within the US military and intelligence community. Alex P. Schmid, "The Definition of Terrorism," in *Routledge Handbook of Terrorism* (New York: Routledge, 2011), 39–99.

52. Derek Gregory reports that new presidential policy guidance on this issue asserts that "males of military age may be non-combatants; it is *not* the

case that all military-aged males in the vicinity of a target are deemed to be combatants." "Dirty Dancing: Drones and Death in the Borderlands," in *Life in the Age of Drone Warfare* (Durham, NC: Duke University Press, 2017), 43.

53. Pugliese, "Drone Casino Mimesis," 508.

54. Ben Anderson, "Facing the Future Enemy: US Counterinsurgency Doctrine and the Pre-insurgent," *Theory Culture & Society* 28, nos. 7–8 (2011): 226.

55. Harwood, "Endless War."

56. Harwood, "Endless War."

57. Harwood, "Endless War."

58. T. Wangsness and J. Franklin, "'Algorithm' and 'Formula,'" *Communications of the ACM* 9, no. 4 (1966): 243.

59. Christian Sandvig, "Seeing the Sort: The Aesthetic and Industrial Defense of 'The Algorithm,'" *Media-N: Journal of the New Media Caucus*, November 22, 2014, http://median.newmediacaucus.org/art-infrastructures-information/seeing-the-sort-the-aesthetic-and-industrial-defense-of-the-algorithm/.

60. Sandvig, "Seeing the Sort."

61. Boris Groys, "The Politics of Installation," *E-Flux* 02 (January 2009), www.e-flux.com/journal/02/68504/politics-of-installation/.

62. Gregory, "The Everywhere War," 239.

63. Harwood, "Endless War."

64. Jean-Louis Baudry, "Ideological Effects of the Cinematographic Apparatus." in *Film Theory and Criticism*, ed. Leo Braudy and Marshall Cohen (Oxford: Oxford University Press, 2004), 293–294.

65. J.L. Austin, *How to Do Things with Words* (Oxford: Oxford University Press, 1975).

66. Jacques Derrida, *The Truth in Painting*, trans. Geoffrey Bennington (Chicago: University of Chicago Press, 1987), 37–-82.

67. While many of the photographs of the installation intentionally or unintentionally reiterate this idea, Ronald Amstutz's beautiful image most forcefully captures it.

68. Gilles Deleuze, "Immanence: A Life," in *Pure Immanence* (New York: Zone Books, 2001), 29.

2. HOW TO PHOTOGRAPH A DRONE

1. Jorge Luis Borges, "On Exactitude in Science," in *Collected Fictions*, trans. Andrew Hurley (New York: Penguin Books, 1999), 325.

2. "US Drone Strikes Kill 28 Unknown People for Every Intended Target, New Reprieve Report Reveals," Reprieve.org, November 25, 2014, https://reprieve.org.uk/press/2014_11_25_us_drone_strikes_kill_28_each_target/. See also Reprieve, "You Never Die Twice: Multiple Kills in the US Drone Program," November 24, 2014, www.reprieve.org/wp-content/uploads/2014_11_24_PUB-You-Never-Die-Twice-Multiple-Kills-in-the-US-Drone-Program-1.pdf.

3. Mike Hill, "Ecologies of War: Dispatch from the Aerial Empire," in *Telemorphosis: Theory in the Era of Climate Change*, vol. 1, ed. Tom Cohen (Ann Arbor, MI: Open Humanities Press, 2012), https://quod.lib.umich.edu/o

/ohp/10539563.0001.001/1:13/--telemorphosis-theory-in-the-era-of-climate-change-vol-1?rgn=div1;view=fulltext.

4. Grégoire Chamayou, *A Theory of the Drone*, trans. Janet Lloyd (New York: New Press, 2015), 74. Because of the large amount of bandwidth required for "live" video, the signal is also often left unencrypted to cut down on this delay. However, this too fails to deliver the kind of shared space-time that the public imagines remote warfare takes place within.

5. Joseph Pugliese, *State Violence and the Execution of Law: Biopolitical Caesura of Torture, Black Sites, Drones* (New York: Routledge, 2013), 186–187. This is similar to what Paul Virilio calls "sightless vision," a "vision without looking—that is itself merely the reproduction of an intense blindness that will become the latest and last form of industrialisation: the industrialisation of the non-gaze." *The Vision Machine*, trans. Julie Rose (Bloomington: Indiana University Press, 1994), 73.

6. Chamayou, *A Theory of the Drone*, 39.

7. Harun Farocki, "Phantom Images," *Public* 29 (2004). Hinting at the strategic function this incompleteness plays within the kill chain, Chamayou also explains: "Imagine you see a shadow resembling a huge dog. If you have access only to the shadow, how can you tell with certainty what object created it? It may simply have been made by an arrangement of someone's hands as part of a shadow play." Chamayou, *A Theory of the Drone*, 51. This gap that opens between signifier and signified does not simply leave the sign open to reconfiguration, but also reverses the trajectory of intelligence. Whereas reconnaissance has historically proceeded from the fragment toward a totality, forming narratives via the gradual accumulation of data, drones reverse this trajectory so that a pregiven totality is brought to bear upon the lone artifact or fragment.

8. Hill, "Ecologies of War."

9. It is significant in this regard that the publication contains not only close-ups of Banner's work but also commissioned images of the city of London by Magnum photographer Paolo Pellegrin.

10. A press release from the Frith Street Gallery in London aptly describes this exchange in predatory terms: "As the drone attempts to focus on the text and images, the downdraught from its rotor blades animates the pages and causes the book to flee before it." Fiona Banner, *Fiona Banner: Font* (digital image, Frith Street Gallery), accessed March 25, 2017, www.frithstreetgallery.com/exhibitions/fiona-banner-font.

11. Banner, *Fiona Banner*.

12. Nadav Assor's *Chariot* and *Ascender/Descender* video loops were taken from a larger work by the artist, also entitled *Lessons on Leaving Your Body*.

13. Assor describes this movement in terms of "an endless loop" in which the drone is both "ascending and descending to and from the heavens." Nadav Assor, "*Chariot* and *Ascender/Descender*" (digital videos and project description, 2014), www.nadassor.net/2014/10/chariot-ascender-descender-2014/.

14. Boris Groys cites Martin Heidegger in describing how technology enacts these transformations: "He shows that historically, the development of technology has been directed towards the decreasing of man's dependence on the accidents to which the natural supply of resources is inevitably prone. One becomes increas-

ingly independent from the sun by storing energy in its different forms—and in general one becomes independent of the annual seasons and the instability of weather. Heidegger does not say this explicitly, but technology is for him primarily the interruption of the flow of time, the production of reservoirs of time in which time ceases to flow towards the future—so that a return to previous moments of time becomes possible." "Art, Technology, and Humanism," *E-Flux* 82 (May 2017), www.e-flux.com/journal/82/127763/art-technology-and-humanism/.

15. Historically, the sun has functioned as both the standard through which linear and cyclical models of time are organized and the means by which one navigates discrete locations within the spatial continuum. Its oscillations between light and darkness, presence and absence are invested in the production of embodied space. This is dramatized by works such as Nancy Holt's *Sun Tunnels* from the 1970s, which presents the daily rituals of the sun as a means of (re-)orienting the body in space. The work materializes solar time within space, intertwining its cyclical ascent and descent with not only spatial reference, but groundedness. The artist explains: "'Time' is not just a mental concept or a mathematical abstraction in the desert. The rocks in the distance are ageless; they have been deposited in layers over hundreds of thousands of year. "Time" takes on a physical presence." Nancy Holt, "Sun Tunnels," in *Theories and Documents of Contemporary Art: A Sourcebook of Artists' Writings*, ed. Peter Selz and Kristen Stiles (Chicago: University of Chicago Press, 2012), 641.

16. Peter Sloterdijk, *Terror from the Air*, trans. Amy Patton and Steve Corcoran (Cambridge. MA: MIT Press, 2009).

17. Svea Bräunert and Meredith Malone, *To See Without Being Seen: Contemporary Art and Drone Warfare* (Chicago: University of Chicago Press, 2016), 14.

18. Derek Gregory, "From a View to a Kill: Drones and Late Modern Warfare," *Theory, Culture and Society*28, no. 88 (2011): 188–215.

19. Gregory, "From a View to a Kill."

20. Bruno Latour, "Spheres and Networks: Two Ways to Reinterpret Globalization" (lecture delivered to Harvard Graduate School of Design, Cambridge, MA, February 17, 2009), www.bruno-latour.fr/sites/default/files/115-SPACE-HARVARD-GB.pdf.

21. Melanie Klein, "A Contribution to the Psychogenesis of Manic-Depressive States," *International Journal of Psycho-Analysis* 16 (1935): 145–174.

22. Such curiosities are the stuff of "click bait," oddities tailor-made for what Martin Hand calls "ubiquitous photography." Martin Hand, *Ubiquitous Photography* (London: Polity, 2012).

23. Joseph Delappe, "*Cowardly Drone*, 2013," www.delappe.net/imaging/mq1-predator-drone---cowardly/.

24. Kris R. Cohen, "What Does the Photoblog Want?," *Media, Culture and Society* 27, no. 6 (2005): 894.

25. Amy Harmon, "Stop Them Before They Shoot Again," *New York Times*, May 5, 2005.

26. In this, the piece maintains a close affinity to what fellow drone artist Trevor Paglen refers to as "entangled photography" or "relational photography." Echoing Nicolas Bourriaud's conception of "relational aesthetics," Paglen

describes a "photography beyond photographs," a mode of practice that foregrounds the politics of image production over the content or finality of the image itself. Seth Curcio, "Seeing Is Believing: An Interview with Trevor Paglen," *Daily Serving*, February 24, 2011, www.dailyserving.com/2011/02/interview-with-trevor-paglen/.

27. Artie Vierkant, "The Image Object Post-Internet," accessed September 30, 2019, http://jstchillin.org/artie/pdf/The_Image_Object_Post-Internet_us.pdf.

28. Stefan Heidenreich, "Freeportism as Style and Ideology: Post-Internet and Speculative Realism, Part I," *E-Flux* 71 (March 2016), www.e-flux.com/journal/71/60521/freeportism-as-style-and-ideology-post-internet-and-speculative-realism-part-i/.

29. As Caren Kaplan argues, this privileging of the elevated view is not simply the product of a need for reliable data, but also serves to reinforce the myth of "virtuous war" (or perhaps "the surgical strike") by way of its remove. "Drone-O-Rama," in *Life in the Age of Drone Warfare*, ed. Lisa Parks (Durham, NC: Duke University Press, 2017), 167.

30. Cited in Ulises Ali Mejias, *Off the Network: Disrupting the Digital World* (Minneapolis: University of Minnesota Press, 2013), 92.

31. Hill, "Ecologies of War.".

32. Derek Gregory, "Theory of the Drone 3: Killing Grounds," *Geographical Imaginations: War, Space and Security*, July 29, 2013, http://geographicalimaginations.com/2013/07/29/theory-of-the-drone-3-killing-grounds/.

33. Gregory, "Theory of the Drone 3."

34. As mentioned in chapter 1, perhaps the most famous example of this reconfiguration is President Barack Obama's 2013 revision to the policy on civilian casualties and drone strikes. Whereas instructions had ordered that operators "ensure" civilians were not targeted, civilian casualties were implicitly allowed by the new policy, which simply specified that they "not be excessive." This was a clear departure from Obama's statement in May of the same year that "before any strike is taken, there must be near-certainty that no civilians will be killed or injured—the highest standard we can set." While it is unknown exactly how these shifts manifested within the implementation of target cycling, a *Washington Post* headline succinctly summarizes the larger consequences of this shift: "Obama War Chiefs Widen Drone Kill Box," December 3, 2013, www.washingtontimes.com/news/2013/dec/3/pentagon-gives-more-room-for-deadly-error-in-drone/.

35. Peters explains: "The history of cloud media, in painting and photography, is the struggle to capture sensuous objects that are also abstract. (The sky was painting abstract impressionist images long before humans did!) Clouds are the original white noise. Well before analog media such as film and sound recording broke the stranglehold of the symbolic in the late nineteenth century, painters struggled to depict cloud colors and forms, often with stunning results. The ability to represent the indefinite is one of the great achievements of modern mathematics and media, and clouds were at the vanguard here too. If you want to understand how meaning works, you have to understand vagueness, and clouds are a chief example." "The Anthropoid Condition: Brían Hanrahan Interviews John Durham Peters," *Los Angeles Review of Books*, July 10, 2015,

http://lareviewofbooks.org/article/the-anthropoid-condition-an-interview-with-john-durham-peters. See also Sean Cubitt, *Finite Media: Environmental Implications of Digital Technologies* (Durham, NC: Duke University Press, 2016).

36. John Durham Peters, *The Marvelous Clouds: Toward a Philosophy of Elemental Media* (Chicago: University of Chicago Press, 2014), 253.

37. Hill, "Ecologies of War."

38. Hill, "Ecologies of War."

39. Seymour M. Hersch, "Rainmaking Is Used as Weapon by U.S.," *New York Times*, July 3, 1972, www.nytimes.com/1972/07/03/archives/rainmaking-is-used-as-weapon-by-us-cloudseeding-in-indochina-is.html.

40. Sloterdijk, *Terror from the Air*.

41. Alexander Abad-Santos, "This 13-Year-Old Is Scared When the Sky Is Blue Because of Our Drones," *Atlantic*, October 29, 2013, www.theatlantic.com/politics/archive/2013/10/saddest-words-congresss-briefing-drone-strikes/354548/. Photojournalist Tomas van Houtryve's series *Blue Sky Days* (2013–2014) takes its name from this same testimony. These works show scenes similar to that which might be seen from the drone's-eye view, but locate them in the United States. Subjects include a wedding, a funeral, and outdoor sports.

42. Rob Nixon, *Slow Violence and the Environmentalism of the Poor* (Cambridge, MA: Harvard University Press, 2013).

43. Lisa Parks, "Vertical Mediation and the U.S. Drone War in the Horn of Africa," in *Life in the Age of Drone Warfare* (Durham, NC: Duke University Press, 2017), 136–137.

44. Parks, "Vertical Mediation and the U.S. Drone War in the Horn of Africa."

45. Some have surmised that this inability of wounds to clot is the product of the use of white phosphorous by Israeli drones. However, there has not been definitive evidence to support this. Raymond Whitaker, "'Tungsten Bombs' Leave Israel's Victims with Mystery Wounds," *Independent*, January 18, 2009, www.independent.co.uk/news/world/middle-east/tungsten-bombs-leave-israels-victims-with-mystery-wounds-1418910.html.

46. Nikhil Mandalaparthy, *UNEP Guide* (Nairobi, Kenya: United Nations Environmental Program, 2015), 19.

47. From Jussi Parikka's "geology of media" to Sean Cubitt's "eco-mediation" of "living media," media studies has come to understand issues of environmental waste and energy consumption not as a byproduct of modern media, but as entangled with the most basic operations of their screens. Cubitt, *Finite Media*; and Jussi Parikka, *A Geology of Media* (Minneapolis: University of Minnesota Press, 2015).

48. Cubitt, *Finite Media*, 4.

49. Mostafa Heddaya, "Landscape Acquired: A Dialogue with Scott Patrick Wiener," Hyperallergic, March 7, 2014, http://hyperallergic.com/113408/landscape-acquired-a-dialogue-with-scott-patrick-wiener/.

50. Heddaya, "Landscape Acquired."

51. "All the events in the Poussin landscape are linked by a common idea, although isolated, set on different planes and separated by large distances."

Cited in David Carrier, *Poussin's Paintings: A Study in Art-historical Methodology* (State College: Penn State Press, 1992), 78.

52. Achille Mbembe, "Necropolitics," *Public Culture* 15, no. 1 (Winter 2003): 28.

53. Mbembe, "Necropolitics," 29.

54. These relations are dramatized in Lawrence Bird's video of the Canadian border, *Parallel 3* (2015). The piece features a seven-hour montage of appropriated images from Google Maps, which simulates movement across the border. In the process, it confirms Bräunert and Malone's claim that "contrary to horizontal wars, vertical wars do not allow for an encounter between opposing sides, because they literally put opponents on different planes." *To See Without Being Seen*, 18.

55. "IWM Talk: Covert War and Colliding Cultures by Mahwish Chishty," Imperial War Museum, December 20, 2016, https://imperialwarmuseum.wordpress.com/2016/12/20/iwm-talk-covert-war-and-colliding-cultures-by-mahwish-chishty/

56. The artist describes the kill box in the following terms: "In weaponry, a Kill Box is a three-dimensional target area, defined to facilitate the integration of coordinated joint weapons fire. It is a joint forces coordination measure enabling air assets to engage surface targets without needing further coordination with commanders and without terminal attack control. The space is defined by an area reference system, but could follow terrain features, be located by grid coordinates or a radius from a center point. A type of Fire Support Coordinating Measure (FSCM), a Kill Box is often defined by a grid reference system—based on lines of Latitude and Longitude - superimposed upon a map of an area of operation. Each square of the grid may be sub-divided into smaller boxes, each of which may carry its own level of permission—or restriction—on the use of air-to-surface or surface-to-surface weapons." "IWM Talk."

57. "IWM Talk."

58. It can even be argued that the arrival of the negative marks the birth of modern photography, as it was only with W. H. Talbot's calotypes that the operation of printing multiple copies of a single image could be realized.

59. In the *Pencil of Nature*, Talbot describes that the calotype system was developed in order to create a fixed and lasting impression of the Lake of Como. However, as Mirjam Brusius points out in an interview, it is clear that the process was also invented to document Assyrian artifacts. Clemency Pleming, "The Victorian Photographer Who Tried to Decipher Assyrian Tablets," *Oxford Arts Blog*, February 5, 2015, www.ox.ac.uk/news/arts-blog/victorian-photographer-who-tried-decipher-assyrian-tablets.

60. This connection persists in contemporary discourse, where not only is it commonplace to describe the relationship between cylinder seals and the tablets they produce in terms of "photographic negatives," but so too is the figure of the wax tablet often called upon in writings on photography in order to convey the medium's unique mode of inscription. Kaja Silverman, *The Miracle of Analogy: The History of Photography Part I* (Stanford, CA: Stanford University Press, 2015), 29, 78. See also William Jerome Harrison, *A History of Photography* (London: Percy Lund and Co., 1897), 19.

3. FROM THE GROUND BELOW

1. The *Living Under Drones: Death, Injury and Trauma to Civilians from US Drone Practices in Pakistan* study was conducted by NYU School of Law and Stanford University Law School. Based on the interviews of hundreds of survivors of drone attacks, the report was published September 25, 2012; it can be found at https://law.stanford.edu/publications/living-under-drones-death-injury-and-trauma-to-civilians-from-us-drone-practices-in-pakistan.

2. Conor Friedersdorf, "Every Person Is Afraid of the Drones: The Strikes' Effect on Life in Pakistan," *Atlantic*, September 25, 2012. www.theatlantic.com/international/archive/2012/09/every-person-is-afraid-of-the-drones-the-strikes-effect-on-life-in-pakistan/262814/

3. Friedersdorf, "Every Person Is Afraid of the Drones."

4. David Rohde, "The Drone Wars," *Reuters Magazine*, January 26, 2012, www.reuters.com/article/us-david-rohde-drone-wars-idUSTRE80P11I20120126.

5. Friedersdorf, "Every Person Is Afraid of the Drones."

6. Matt Delmont, "Drone Encounters: Noor Behram, Omer Fast, and Visual Critiques of Drone Warfare" *American Quarterly* 65, no. 1 (March 2013): 193–202.

7. Grégoire Chamayou and Janet Lloyd, *A Theory of the Drone* (New York: New Press, 2015), 118.

8. Sigmund Freud, "Inhibitions, Symptoms and Anxiety," *Standard Edition*, Vol. 20.

9. Carl Schmitt, *The Concept of the Political*, trans. George Schwab (Chicago: University of Chicago Press, 2007), 26.

10. Alexander Galloway and Eugene Thacker, *The Exploit: A Theory of Networks* (Minneapolis: University of Minnesota Press, 2007), 65.

11. Galloway and Thacker, *The Exploit*, 63.

12. Processes that utilize this additive mode of construction include selective laser melting (SLM) or direct metal laser sintering (DMLS), selective laser sintering (SLS), and fused deposition modeling (FDM).

13. Gilles Deleuze, *The Fold: Leibniz and the Baroque* (London: Athlone Press, 1993), 20.

14. Chris Coker, *Future War* (Cambridge, UK: Polity Press, 2015), 105.

15. Accompanying booklet for Gonçalo F. Cardoso and Ruben Pater's *A Study in 21st Century Drone Acoustics* (2015).

16. There had been numerous recordings of birds on wax cylinders going as far back as the nineteenth century, but these works featured captive animals rather than birds in their natural environment. While the results of this initial attempt by Allen were mixed, the experience opened the door to the possibility of relatively high fidelity recordings of wild birds in situ, an untapped genre that would prove invaluable as a teaching tool. Relying on trucks (or mules when roads were not available), the system had obvious limitations and was soon replaced by more familiar sound recording devices such as the Nagra, which could be thrown over the shoulder and carried by one person. In fact, it was Kellogg himself who in the 1940s designed the first prototype for a truly portable recorder.

17. It is noteworthy in this respect that the first recordings of bird songs made during the nineteenth century were themselves comprised of human imitations.

18. Judith Butler, "Ruled Out: Vocabularies of the Censor," in *Censorship and Silencing: Practices of Cultural Regulation*, ed. Robert C. Post (Los Angeles: Getty Research Institute, 1998), 249.

19. Gilles Deleuze and Felix Guattari, *A Thousand Plateaus: Capitalism and Schizophrenia*, trans. Brian Massumi (Minneapolis: University of Minnesota Press, 1987), 95.

20.

21. Einar Sneve Martinussen, Jørn Knutsen, and Timo Arnall, "Satellite Lamps," *Kairos: A Journal of Rhetoric, Technology and Pedagogy* 19, no. 1 (Fall 2014), http://kairos.technorhetoric.net/19.1/inventio/martinussen-et-al/index.html.

22. James Bridle, "The Art of Instagram: Dronestagram," Tate Museum, *Blogs and Channel*, May 8, 2015, www.tate.org.uk/context-comment/articles /art-of-instagram-dronestagram (link no longer active in September 2019).

23. While live video stream is typically transmitted through a microwave signal called the common data link (CDL), drones transmit control commands via encrypted satellite communications.

24. Christina Bonnington and Spencer Ackerman, "Apple Rejects App That Tracks US Drone Strikes," *Wired*, August 30, 2012, www.wired.com/danger room/2012/08/drone-app/.

25. Karlis Kalnins first proposed the term *locative media* during the Art and Communication Festival, May 16–17, 2003, in Riga, Latvia.

26. Tim Lenoir, Foreword to *New Philosophy for New Media* by Marc Hansen (Cambridge, MA: MIT Press, 2004), xiii.

27. As Didem Özkul describes, this bias also informs practices of location sharing, as users "retrieve or disclose locational information in everyday life [in order to] . . . deal with the anxiety of getting lost, organize daily activities and be punctual, gain a feeling of security, discover new places, establish social relations, and maintain close ties, among other things." "Location as a Sense of Place," in *Mobility and Locative Media: Mobile Communication in Hybrid Spaces*, ed. Adriana de Souza e Silva and Mimi Sheller (New York: Routledge, 2014), 102.

28. Terri Rueb, "Restless: Locative Media as Generative Displacement," in *Mobility and Locative Media*, ed. de Souza e Silva and Sheller (New York: Routledge, 2014).

29. Rueb, "Restless."

30. Mimi Sheller, "Mobile Art: Out of Your Pocket," in *Routledge Companion to Mobile Media*, ed. Gerard Goggin and Larissa Hjorth (New York: Routledge, 2014), 199.

31. Ingrid Richardson, "Mobile Technosoma: Some Phenomenological Reflections on Itinerant Media Devices," *FibreCulture Journal* 6 (2005), http:// six.fibreculturejournal.org/fcj-032-mobile-technosoma-some-phenomenologi cal-reflections-on-itinerant-media-devices/.

32. Richardson, "Mobile Technosoma."

33. L. Y. Hsu, T. Lepetit, and B. Kante, "Extremely Thin Dielectric Metasurface for Carpet Cloaking," *PIER: Progress in Electromagnetic Research* 152 (2015): 33–40.

34. David Hambling, "Cloak of Light Makes Drone Invisible?," *Wired*, May 9, 2008, www.wired.com/2008/05/invisible-drone/; and Yasmin Tayag, "Drones Can Be Rendered Invisible with New Teflon Cloaking Technology: By Reflecting their Surroundings, UAVs Can Disappear," *Inverse Innovation*, September 8, 2015, www.inverse.com/article/5975-drones-can-be-rendered-invisible-with-new-teflon-cloaking-technology.

35. David Hambling, "Silence of the Drones: How to Quiet That Annoying Aerial Buzz," *New Scientist*, June 10, 2015, www.newscientist.com/article/dn27696-silence-of-the-drones-how-to-quiet-that-annoying-aerial-buzz/.

36. Steve Henn, "If There's Privacy in the Digital Age, It Has a New Definition," NPR, *All Tech Considered: Policy* (blog), March 3, 2014, www.npr.org/sections/alltechconsidered/2014/03/03/285334820/if-theres-privacy-in-the-digital-age-it-has-a-new-definition.

37. Hal Abelson, "Seductive Myths about Privacy," July 7, 2010, www.w3.org/2010/policy-ws/slides/09-Abelson-MIT.pdf

38. Cora Currier, "How the Gov't Talks about a Drone Program It Won't Acknowledge Exists," *The Drone War* (blog), *ProPublica*, September 13, 2012. www.propublica.org/article/how-the-govt-talks-about-a-drone-program-it-wont-acknowledge.

39. David Kravets, "CIA Refuses to Acknowledge Drone Targeted Killings," *Wired*, August 13, 2013, www.wired.com/2013/08/drone-targeted-killings.

40. Glenn Greenwald, "ACLU Sues Obama Administration over Assassination Secrecy" *Salon*, February 2, 2012, www.salon.com/2012/02/02/aclu_sues_obama_administration_over_assassination_secrecy/; and Glenn Greenwald, "Obama DOJ Again Refuses to Tell a Court Whether CIA Drone Program Even Exists," *Guardian*, February 14, 2013, www.theguardian.com/commentisfree/2013/feb/14/cia-aclu-drone-secrecy.

41. Michael Calderone, "Robert Gibbs Told Not to Acknowledge Drone Program Exists as White House Press Secretary," *Huffington Post*, February 24, 2013, www.huffingtonpost.com/2013/02/24/robert-gibbs-drones-white-house_n_2753183.html.

42. Ian Almond, "Derrida and the Secret of the Non-Secret: On Re-Spirtualising the Profane," *Literature & Theology* 17. no. 4 (December 2003): 461.

43. Almond, "Derrida and the Secret of the Non-Secret."

44. Jacques Derrida, "How to Avoid Speaking: Denials," in *Derrida and Negative Theology*, ed. edited by Harold Coward and Toby Foshay (Albany: State University of New York Press, 1992): 95.

45. Slavoj Žižek, *The Sublime Object of Ideology* (New York: Verso, 1989).

46. Sigmund Freud, "Fetishism," in *The Complete Works of Sigmund Freud*, vol. 21 (London: Hogarth, 1927), 157.

47. Robert Pfaller, "Little Gestures of Disappearance: Interpassivity and the Theory of Ritual," *European Journal of Psychoanalysis* 16 (2003).

48. Robert Pfaller, *Interpassivity: The Aesthetics of Delegated Enjoyment* (Edinburgh: Edinburg University Press, 2017).

49. Pfaller, "Little Gestures of Disappearance."

50. Pfaller, "Little Gestures of Disappearance.".

51. The collectivizing nature of these relations, their ability to produce publics across bodies, dramatizes what Dominic Pettman refers to as a kind of nostalgic return to the "will-to-synchronize." He explains: "Those readers old enough to remember what is was like to live before the Internet will recall the strange phenomenon where the general noosphere seduced us by its sheer beckoning presence. Thus, we find ourselves listening to terrible songs or talk shows on the radio in the car rather than listening to our perfectly sequenced mixtape or intriguing audiobook. Or we would end up flicking from channel to channel on the TC, preferring this cathodic wasteland to the stack of quality VHS videos that sat in the corner of the living room. Such perverse behavior exhibits a profound and tenacious will-to-synchronize." Working from this scenario it becomes clear that while the kill chain operates via the displacement of agency (and culpability), it also provides the consolidation of agency necessary for the unique relations of asymmetry of its operations. In this regard, it provides the means of not only sustaining national identity in the globalized networks of drone power, but also of validating and collectivizing the relations of the non-secret through its operations. Dominic Pettman, *Infinite Distraction* (New York: Polity, 2015), 51.

4. THE ANIMAL REMAINDER

1. See, for example, *Animals and War: Confronting the Military-Animal Industrial Complex,* edited by Anthony J. Nocella II, Colin Salter, and Judy K. C. Bentley (Lanham, MD: Lexington Books, 2013); and Jill Cooper, *Animals in War* (London: Corgi, 1983).

2. Sigmund Freud, *The Standard Edition of the Complete Psychological Works of Sigmund Freud*, ed. James Strachey, vol. 21, *1927–1931* (New York: Norton, 1961), 39.

3. This formation echoes a foundational relationship between modernity and the animal. As the latter disappeared from daily life at the hands of the former, the animal subsequently became a mainstay in discourses such as industry, psychoanalysis, and quantum physics. As Akira Lippit puts it, "technological instruments and media [of the later nineteenth century] began to serve as virtual shelters for displaced animals. . . . [T]echnology and ultimately cinema came to determine a vast mausoleum for animal being" (187). However, reading this relationship of the animal remainder diachronically opens up the prospect of a disembodied animality that is not simply the "emblem" of modernity, but rather an enduring agency that continues to exert influence upon the apparatus long after the animal body has been left behind by successive technologies. Akira Lippit, *Electric Animal: Towards a Rhetoric of Wildlife* (Minneapolis: University of Minnesota Press, 2008).

4. Donna Jeanne Haraway, *When Species Meet* (Minneapolis: University of Minnesota Press, 2008), 261.

5. The record for the most words carried during a single flight was thirty thousand. Frank W. Lane, "Pigeons as Birds of War," *Flight: The Aircraft and Engineer* (October 21, 1943): 450.

6. Lane, "Pigeons as Birds of War."

7. As Peter Adey points out, the arrival of this notion of "seeing from the air" marked the beginning of "imperial exploration, colonial administration and development." *Aerial Life: Spaces, Mobility, Affect* (New York: Wiley, 2010), 87.

8. While Neubronner's images are not directly referenced in this particular text, the lineage implied by this montage relates to the central argument of the essay which is that while the mind articulates a seemingly infinite realm of possibilities, this labor always transcends the technological capacity in which it takes place. In trying to bring this mode of vision into being, technology undergoes what Moholy-Nagy calls "detours," which are opposed to "directness" of the mind, i.e., its ability to envision worlds abstractly and without impediment. "Directness of the Mind; Detours of Technology," trans. Sibyl Moholy-Nagy, in *Moholy-Nagy*, ed. Richard Kostelanetz (New York: Praeger, 1970).

9. Franz Marc, "How Does a Horse See the World," in *Theories of Modern Art: A Source Book by Artists and Critics*, ed. Herschel Browning Chipp and Peter Howard Selz (Berkeley: University of California Press, 1968), 178.

10. Marc to Maria, April 12, 1915, in *Letters from the War: Franz Marc*, ed. Klaus Lankheit and Uwe Steffen (Bern, Switzerland: Peter Lang, 1992), 46.

11. As evolutionary theory and psychoanalysis consistently positioned the human in relation to the animal, modernism often conjured a fantastic hybridity as a means of disclosing the irrational base of the psyche or the primitive roots of the biology of the human. To this extent, the interpenetration of machine and animal that characterized Neubronner's pigeon photographers would appear to fall in line with F.T. Marinetti's "snorting machines . . . puffing on the rails like enormous steel horses with long tubes for bridle," or the chronophotographic legs of Giacomo Balla's dog in *Dynamism of a Dog on Leash* painting (1912), or even George Grosz's paintings of war veterans, whose arms and legs take on the mechanization of the war. F.T. Marinetti, "The Founding and Manifesto of Futurism," in *Futurism: An Anthology*, ed. Lawrence Rainey, Christine Poggi, and Laura Wittman (New Haven, CT: Yale University Press, 2009), 49.

12. There is some debate about how the pigeon stitches together phenomena that move between these zones. One of the most interesting hypotheses is that the bird's constant head motions allow it to create continuity between these systems. See, for example, Stephen E.G. Lea and Winand H. Dittrich, "What Do Birds See in Moving Video Images?," in *Picture Perception in Animals*, ed. Joel Fagot (Philadelphia, PA: Taylor and Francis, 2000), 148–150.

13. Ian P. Howard and Brian J. Rogers, *Binocular Vision and Stereopsis* (Oxford: Oxford University Press, 1995), 645.

14. See Marshall McLuhan, "The Gadget Lover: Narcissus as Narcosis," in *Understanding Media: The Extensions of Man*, ed. Terrence W. Gordon (Corte Madero, CA: Gingko Press, 1994), 63–70.

15. Ruth Erikson, "Animal-Borne: Embodied Point of View and the Ethics of Identification," in *Cine-Ethics: Ethical Dimensions of Film Theory, Practice and Spectatorship* (New York: Routledge, 2014), 214–231.

16. For the last fifty years, the navy has enlisted dolphins to detect and disarm underwater mines. The bio-sonar system of these animals utilizes echolocation to detect prey and predators in low or no visibility conditions. It is so accurate that

it can tell the difference between a golf ball and a ping pong ball based on density. In the military context, dolphins are outfitted with a locator ping to alert operators to locations of mines, even those that may be underwater. In addition, cameras are attached to the dorsal fin or held in the dolphin's mouth in order to conduct surveillance operations of oncoming vehicles and clue divers into what kind of mine they have encountered. The navy is currently retiring its dolphin program as a new underwater drone called the Knife Fish comes into use. The new technology will use a special low frequency radar system, which will reference a database of profiles to compare readings against a catalog of previous impressions. In this way, mine identification algorithms will learn this indexical language, which the dolphin intuits from vibrations received in its jawbone. Dolphins also use "bubble nets" to distinguish between targets and clutter in the water. This "bubble nets" method is the basis for a new sonar system being developed called twin inverted pulse radar (TWIPR). In these instances, the dolphin sensorium is initially co-opted and then formalized via an interface that structures human interaction with the technology. See T.G. Leighton, G.H. Chua, P.R. White, K.F. Tong, H.D. Griffiths, and D.J. Daniels, "Radar Clutter Suppression and Target Discrimination Using Twin Inverted Pulses," *Proceedings of the Royal Academy A* 469, no. 2160 (December 8, 2013), https://doi.org/10.1098/rspa.2013.0512. For a summary, see "Dolphins Inspire New Radar System," UCL Department of Electronic and Electrical Engineering, accessed March 14, 2015, https://www.ucl.ac.uk/news/2013/oct/dolphins-inspire-new-radar-system.

17. Haraway, *When Species Meet*, 262.

18. Beatriz da Costa, "Pigeonblog 2006–08," *Beatriz da Costa's Blog & Project Hub*, accessed February 12, 2015, http://nideffer.net/shaniweb/pigeonblog.php.

19. Beatriz da Costa, with Cina Hazegh and Kevin Ponto, "Interspecies Coproduction in the Pursuit of Resistant Action," accessed February 12, 2015, http://nideffer.net/shaniweb/files/pigeonstatement.pdf.

20. Pigeon Blog Project Statement, *Beatriz da Costa's Blog and Project Hub*, http://nideffer.net/shaniweb/files/pigeonstatement.pdf.

21. In addition to these references, the film also features a shot of a computer screen that shows an ad for Little Red Riding Hood in Facebook. The young woman also wears a red bikini.

22. Alan Dundes, "Interpreting 'Little Red Riding Hood' Psychoanalytically," in *Little Red Riding Hood: A Case Book* (Madison: University of Wisconsin Press, 1989), 206–207.

23. Michel Chion, *Audio-Vision: Sound on Screen*, trans. Claudia Gorbman (New York: Columbia University Press, 1994), 9.

24. Nasser Hussain, "The Sound of Terror: Phenomenology of a Drone Strike," *Boston Review*, October 16, 2013, http://bostonreview.net/world/hussain-drone-phenomenology.

25. As mentioned previously, the constant buzzing sound of the drone has been linked to anticipatory anxiety, insomnia, and even psychosis. See C. Christian Fair, "Ethical and Methodological Issues in Assessing Drones' Civilian Impacts in Pakistan," *Washington Post*, October 6, 2014, www.washingtonpost.com/blogs/monkey-cage/wp/2014/10/06/ethical-and-methodological-issues-in

-assessing-drones-civilian-impacts-in-pakistan/; and "Drones Impact on Mental Health in Pakistan," *The Borgen Project* (blog), December 31, 2013, http://borgenproject.org/drones-impact-mental-health-pakistan/.

26. Of the sensory channels drones employ in forging this mode of address, sound is especially important. Not only does it often take precedence over the visual as the primary identifier of the UAV for those on the ground, but so does its semantic operation perform the self-revelatory logic of drone power in reflexive fashion. The Pakistani word for drones is *bangana*, an onomatopoeia for the unnerving metallic hum its engines make. This relationship, whereby the sound an entity makes serves as its name, is familiar to bird-watchers, whose first lesson in the analysis of avian song often centers on species that "say their own name." Like the drone, the cuckoo, the bobwhite, and the killdeer are not simply rendered legible or identifiable by their song, but are given a primary power of naming in the process. This name is not descriptive or even anaphoric, but rather is produced through a self-referential sonic loop that always already ropes nonsymbolic sound into speech. Accordingly, when drones declare their own presence to Pakistanis sonically, they do so in a literal manner, such that the sky itself calls out "drone!" In actively announcing themselves in this way, drones enact a mode of interpolation that has more in common with Althusser's police officer than with Foucault's warden. While the prisoners of Bentham's prison were never certain if the warden was present, subjects of the drone's eye not only know they are there, but become subjects by virtue of this knowing.

27. As John Durham Peters states: "[S]ounds often provide cues about where to look, but sights rarely tell us where to listen." *The Marvelous Clouds: Toward a Philosophy of Elemental Media* (Chicago: University of Chicago Press, 2014), 302. "The light of god" exemplifies this scenario. The phrase refers to a laser beam that marks an impending target and is only visible to US troops with night vision goggles.

28. Jacques Ranciere, *The Politics of Aesthetics: The Distribution of the Sensible* (New York: Continuum, 2004).

29. Lippit, *Electric Animal*, 166.

30. Lippit, *Electric Animal*, 166, 170.

31. One finds evidence of this relation in the peculiar mash-up that adorns the cover of the press release for Festo's 2011 "smart bird" vehicle. Modeled after the seagull in appearance and operation, the camera-equipped drone is shown soaring through blue skies as an actual bird of the same species sits on a nearby cliff, staring at the creation with a perplexed look of wonder. "Smartbird: Bird Flight Deciphered," Festo, accessed March 2, 2016, www.festo.com/net/SupportPortal/Files/46270/Brosch_SmartBird_en_8s_RZ_110311_lo.pdf.

32. Étienne Balibar, "Racism and Nationalism," in *Race, Nation, Class: Ambiguous Identities*, trans. Chris Turner (New York: Verso, 1988), 57.

33. Known as the "Maywand District Murders," at least three Afghani civilians were killed by US Army soldiers for sport in 2010. The soldiers referred to themselves as the "kill team."

34. James J. Weingartner, "Trophies of War: U.S. Troops and the Mutilation of Japanese War Dead, 1941–1945," *Pacific Historical Review* 61, no. 1 (February 1992): 53–67.

35. Michael Hastings, "The Rise of the Killer Drones: How America Goes to War in Secret," *Rolling Stone*, April 16, 2012, www.rollingstone.com/politics/news/the-rise-of-the-killer-drones-how-america-goes-to-war-in-secret-20120416.

36. See Grégoire Chamayou, "Psychopathologies of the Drone," in *A Theory of the Drone*, trans. Janet Lloyd (New York: New Press, 2015), 106–113.

37. The relevance of these relations of the windshield to the computer screen is confirmed not only by the shared immobility of the operator in both instances, but also by the technical migration of the former into the latter. See Dana Hull and Ian King, "The Car Windshield Is Turning into a Computer Screen," *Bloomberg*, January 22, 2015, www.bloomberg.com/news/articles/2015-01-22/the-car-windshield-is-turning-into-a-computer-screen.

38. Paul Virilio, "Dromoscopy, or the Ecstasy of Enormities," *Wide Angle* 20, no. 3 (1998): 11–22 (emphasis added).

39. Theodor Adorno, *Minima Moralia: Reflections on a Damaged Life* (New York: Verso, 2006) 19, 20.

40. Gregory S. McNeal, "Are Targeted Killings Unlawful? A Case Study in Empirical Claims Without Empirical Evidence," in *Targeted Killings: Law and Morality in an Asymmetrical World*, ed. Claire Finkelstein, Jens David Ohlin, and Andrew Altman (Oxford: Oxford University Press, 2012), 326–346.

41. Virilio, "Dromoscopy, or the Ecstasy of Enormities," 21. Since Virilio's diagnosis forty years ago, not only has the windshield come to merge with the computer screen via "head-up displays" (HUDs) of contemporary automobiles, but so has the screen itself reproduced this depthless experience of "speed pictures" in the form of racing simulators and first person shooter games. Originally developed to keep the eyes of fighter pilots on the skies rather than the gauges and instruments of the cockpit, HUDs project information onto the actual windshield of the car. See Hull and King, "The Car Windshield Is Turning into a Computer Screen."

42. Hull and King, "The Car Windshield Is Turning into a Computer Screen."

43. Chris Woods, *Sudden Justice: America's Secret Drone Wars* (Oxford: Oxford University Press, 2015), 150.

44. Lisa Parks, "Vertical Mediation and the US Drone War in the Horn of Africa," in *Life in the Age of Drone Warfare*, ed. Caren Kaplan (Durham, NC: Duke University Press, 2017),145.

45. Joseph Pugliese, "Drone Casino Mimesis: Telewarfare and Civil Militarization," *Journal of Sociology* 52, no. 3 (2016): 500–521.

46. Étienne Balibar, "Racism and Nationalism," in *Race, Nation, Class: Ambiguous Identities*, trans. Chris Turner (New York: Verso, 1988).

47. Jordan Crandall, "Ontologies of the Wayward Drone: A Salvage Operation," *CTheory* (February 11, 2011), http://ctheory.net/ctheory_wp/ontologies-of-the-wayward-drone-a-salvage-operation/. The growing presence of drones in civilian contexts has made such events increasingly common. For example, a drone pilot accidently flew his vehicle into the fortieth floor of the Empire State building in New York City in February 2016. In May of the same year, a UAV crashed through a Capitol Hill home, its camera still recording as the couple inside picked up the DJI Phantom from their living room floor. In 2017 a similar

incident occurred at the Space Needle in Seattle when a drone plummeted onto the top platform of the structure, stunning a group of workers who were putting up holiday lights.

48. The members of the group are Ali Rez, Saks Afridi, Assam Khalid, Akash Goel, Insiya Syed, Noor Behram, Jamil Akhtar, and the JR InsideOut project. The idea grew out of the *Inside Out* project, in which large-scale images of local people related to a given issue or cause were pasted on city streets. It began to gain steam when the artists joined with the UK legal group Reprieve, which has been working to expose the cost of drone strikes in Pakistan for some time. Jens Eder and Charlotte Klonk, *Image Operations: Visual Media and Political Conflict* (Manchester, UK: Manchester University Press, 2016). See also Sophia Saifi, "Not a 'Bug Splat:' Artists Give Drone Victims a Face in Pakistan," CNN, April 9, 2014, www.cnn.com/2014/04/09/world/asia/pakistan-drones-not-a-bug-splat/.

49. The vinyl sheet was eventually recycled as roofing material by local residents. Alice Speri, "Giant Art Installation in Pakistan Tells US Drone Operators People Aren't 'Bug Splat,'" *Vice News*, April 7, 2014, https://news.vice.com/article/giant-art-installation-in-pakistan-tells-us-drone-operators-people-arent-bug-splat.

50. "About" page, *#NotABugSplat* website, accessed December 20, 2016, https://notabugsplat.com/about/.

51. "The face is exposed, menaced, as if inviting us to an act of violence. At the same time, the face is what forbids us to kill." Emmanuel Levinas, *Ethics and Infinity: Conversations with Philipe Nemo*, trans. Richard A. Cohen (Pittsburgh: PA: Duquesne University, 1995), 86.

52. The face in Levinas's formulation contains clear references to God and the theological. He states that "in the face such as I describe its approach, is produced the same exceeding of the act by that to which it leads. In the access to the face there is certainly also an access to the idea of God." Levinas, *Ethics and Infinity*, 92.

53. Levinas, *Ethics and Infinity*, 89.

54. Chamayou and Lloyd, *Theory of the Drone*, 117–118.

55. Svea Bräunert and Meredith Malone, *To See Without Being Seen: Contemporary Art and Drone Warfare* (Chicago: University of Chicago Press, 2016), 18–19.

56. Simone Bignall, "Dismantling the Face: Pluralism and the Politics of Recognition," *Deleuze Studies* 6, no. 3 (2002): 395.

57. Bignall explains that "facialisation relies upon the interpretive work of the subject, who is responsible for the 'territorialisation'—the selection and establishment—of a privileged set of significations from a range of possible alternatives, and the consolidation of these meanings in accordance with a given mode of expression. In this sense, the face is a 'territory' that is carved out from a broader 'landscape'—the collection of diverse orders of meaning that form any given social, political and discursive milieu. The face expresses a dominant outlook and mode of understanding: it encodes a particularly powerful regime of signs." "Dismantling the Face," 395.

58. Wendy Kozol, *Distant Wars Visible: The Ambivalence of Witnessing* (Minneapolis: University of Minnesota Press, 2014), 199, 203.

59. Gilles Deleuze and Félix Guattari, *A Thousand Plateaus: Capitalism and Schizophrenia*, trans. Brian Massumi (Minneapolis: University of Minnesota, 1987), 187.

60. This is exemplified by the work of Francis Bacon, which Deleuze describes as painting heads rather than faces.

61. Maurice Blanchot, *The Instant of My Death* (Stanford, CA: Stanford University Press, 1998), 3.

62. Jean-Luc Nancy, *Dis-Enclosure: The Deconstruction of Christianity*, trans. Bettina Bergo, Michael Bradley Smith, and Gabriel Malenfant (New York: Fordham University Press, 2009), 89.

63. Jennifer Yusin, "Writing the Disaster: Testimony and *The Instant of My Death*," *Colloquy: Text, Theory, Critique* 10 (November 2015): 137.

64. Maurice Blanchot, *The Writing of Disaster* (Omaha: University of Nebraska Press, 2015), 71–72.

65. Mike Pearl, "The #NotaBugSplat Art Piece in Pakistan Won't Be Making Drone Pilots Feel Empathy," *Vice News*, April 8, 2014, www.vice.com/en_uk/article/this-giant-art-piece-in-pakistan-wont-be-making-drone-pilots-feel-empathy.

66. Tom Holert, "Sensorship: The Seen Unseen of Drone Warfare," in *Image Operations: Visual Media and Political Context*, ed. Jen Elder and Charlotte Klonk (Manchester, UK: Manchester University Press, 2017), 112.

67. Kate Drazner Hoyt, "Ethics of Network Subjectivity," *Technoculture* 4 (2014), https://tcjournal.org/vol4/hoyt.

68. Hoyt, "Ethics of Network Subjectivity."

69. As David L. Clarke explains: "Robbing the prisoners of the power to speak, the Nazis cause them to question their ability to reason—language and thinking being the exemplary characteristics by which the human has always been decisively distinguished from the animal. What breaks the binding force of this animalization is an animal, 'Bobby.' Wandering into the camp, the dog "unwittingly" bears witness to the humanity of Levinas and the other prisoners, remembering what the Nazis, in their unremitting savagery, have forgotten. Like some strange, reversed pharmakos, Bobby is cast into (not out of) the mock-polis of the camp, restoring it – albeit momentarily—to a semblance of ethical 'health.'" "On Being 'the Last Kantian in Nazi Germany': Dwelling with Animals after Levinas," in *Animal Acts: Configuring the Human in Western History*, ed. Jennifer Ham and Matthew Senior (New York: Routledge, 1997), 165–198.

70. Peters, *Marvelous Clouds*, 371.

71. Original quote by Max Ernst, cited in Pelagia Goulimari, *Literary Criticism and Theory: From Plato to Postcolonialism* (New York: Routledge, 2014).

72. While not necessarily clearing the artist of ethical considerations, it is important to point out that the animals in all of Jensen's projects are already deceased before they are integrated into his work.

73. Lane, "Pigeons as Birds of War," 449.

74. Peters, *Marvelous Clouds*, 75.

5. SHOWING SENSING

1. Rosalind Krauss, *A Voyage on the North Sea: Art in the Age of the Post-Medium Condition* (London: Thames & Hudson, 1999), 64.

2. Krauss, *A Voyage on the North Sea*, 64.

3. See Introduction to "And Phenomenology of the End: Cognition and Sensibility in Transition from Conjunctive to Connective Mode of Social Communication" (PhD diss., Aalto University, 2014).

4. Quoted in Patrice Pavis, "The State of Current Theatre Research," *Applied Semiotics/Sémiotique Appliquée* 1, no. 3 (1997): 203.

5. Gilles Deleuze, "Postscript on Societies for Control," *October* 59 (Winter 1992): 4.

6. Roland Barthes, "Leaving the Movie Theater," in *The Rustle of Language* (New York: Hill and Wang, 1986).

7. Aaron Sankin, "The Loneliness of the Long Distance Drone Pilot," *Kernel*, February1, 2015, http://kernelmag.dailydot.com/issue-sections/headline-story/11553/air-force-drone-pilot-problem/.

8. Paolo Virno, "PostFordist Semblance," *SubStance* 36, no. 1 (2007): 45. This was confirmed in 2011 when a drone station at Creech Air Force Base was infected with a credentials-stealing virus. In the official explanation provided by General Robert Kehler, it was noted that the virus entered the system "from the wild," that is, outside of the parameters of the military's para-network. According to media reports, all signs seemed to suggest that the most likely source of the virus was the online game *Mafia Wars*, which pilots had apparently played during or between drone flights. Rather than being a simple infraction, the incident can be seen as a symptom of the porosity of the Conex box and the way in which its partitioning off from the outside enacts processes of enfolding that resonate with the structure of cognitive labor. Noah Shachtman, "Military 'Not Quite Sure' How Drone Cockpits Got Infected," *Wired*, October 19, 2011, www.wired.com/2011/10/military-not-quite-sure-how-drone-cockpits-got-infected/.

9. Through this slippage, these scenes produce lost time in a dual sense, in that their appearance marks dead spots in the narrative where attention runs aground, and at the same time, these suspensions mark the death of an onscreen subject.

10. Garrett Stewart, "Digital Fatigue: Imaging War in Recent American Film," *Film Quarterly* 62, no. 4 (Summer 2009): 45–55.

11. Anthony Lewis, "Distant Emotions: 'Good Kill" and 'Slow West,'" *New Yorker*, May 18, 2015, www.newyorker.com/magazine/2015/05/18/distant-emotions.

12. The site for these scenes of escape is the Arizona desert. Separating the base from civilian life, the undifferentiated terrain of the desert forms a virtual field of nonspecificity, which Egan's car pierces in cathartic fashion. Ethan Hawke himself proposed the addition of the car to the script. He was inspired by the 1980s fighter pilot blockbuster *Top Gun* (1986), in which intense affective sequences of flight punctuate masculine posturing. As warfare still meshed with codes of identification and continuity cinema, the F-16 served as the ultimate tool of misrecognition in the 1980s film, an absurdly efficient convergence point for male virility, desirability, and heroism. Fueling this interpenetration is a con-

catenation of screens that include the muscle car, the F-16, and the filmic frame. In the film, the migration of these affective sequences across screens serves to multiply their intensity such that the pervasive displacement of motion to the screen bolsters the spectator's investment in the filmic illusion. However, within the windowless air-conditioned units from which drone warfare is waged, these references to *Top Gun* only further dramatize the impotence of the operator and the nostalgia for the embodiment of combat. In the Conex box, the networking of screens dissipates rather than amplifies the agency of the operator.

13. In this regard *Good Kill* echoes a recurring theme of the film *National Bird* (2016), which catalogs the effects of drone warfare from the point of view of drone operators and Afghan victims. "Imagine if this was happening to us," Lisa, a whistleblower from the US Air Force, asks in the film, "Would parents still allow kids out to play if they didn't know if today is the day that something is going to fall out of the sky and kill them?"

14. The duality formed by these conflicting propositions echoes what Alan Liu describes as the contradictory position of networks in the cultural imaginary: "If the network is our contemporary intuition of infinity, then its boundlessness is match by an equally infinite, equally unreal hunger for security." Seeing as though the representation of connectivity in our post–Cold War milieu seems destined to summon these opposing associations of infinite expansion and absolute insecurity, it would seem only logical that this irreconcilability be magnified in the context of terror and martial violence. Nonetheless, *Good Kill*'s refusal to resolve these readings has frustrated viewers who crave a definitive denunciation of drone violence. Alan Liu, *The Laws of Cool: Knowledge Work and the Culture of Information* (Chicago: University of Chicago Press, 2004), 42.

15. Gene Siskel, "The Touch That Transcends Violence and Death," *Chicago Tribune*, November 11, 1973.

16. Garrett Stewart, *Closed Circuits: Screening Narrative Surveillance* (Chicago: University of Chicago Pres, 2014), 172.

17. D.N. Rodowick, *The Virtual Life of Film* (Cambridge, MA: Harvard University Press, 2007).

18. See Rudolph Arnheim, *Film as Art* (London: Faber and Faber, 1933); and Béla Balázs, *Theory of Film: Character and Growth of a New Art*, trans. Edith Bone (London: Dennis Dobson Ltd., 1952), 24–32.

19. In Barthes's essay, the drama of light illuminating partial bodies in the seats in front of the spectator dislodged the film from the screen so that it spilled into the built environment, becoming a mongrel medium bound up with queer desire. Barthes, "Leaving the Movie Theater." See also Guo-Juin Hong, "Theatrics of Cruising: Bath Houses and Movie Houses in Tsai Ming-liang's Films," in *Queer Sinophone Cultures*, ed. Howard Chiang and Ari Larissa Heinrich (New York: Routledge, 2013); and Nicholas de Villiers, "Leaving the Cinema: Metacinematic Cruising in Tsai Ming-liang's Goodbye, Dragon Inn," *Jump Cut* 50 (Spring 2008), www.ejumpcut.org/archive/jc50.2008/DragonInn/.

20. Chiel Kattenbelt, "Intermediality in Theatre and Performance: Definitions, Perceptions and Medial Relationships," *Culture, Language and Representation* 6 (2008): 22.

21. Kattenbelt, "Intermediality in Theatre and Performance," 23.
22. George Brant, *Grounded* (London: Oberon Books, 2013).
23. Quoted in Pavis, "The State of Current Theatre Research," 203.
24. Heinrich Müller, *Rotwelsch* (Berlin: Merve, 1984), 38.
25. Suzanne M. Jaeger, "Embodiment and Presence: The Ontology of Presence Reconsidered," in *Staging Philosophy: Intersections of Theater, Performance and Philosophy*, ed. David Krasner and David Z. Saltz (Ann Arbor: University of Michigan Press, 2006), 122.
26. T.S. Eliot, "The Hollow Men," in *Selected Poems* (London: Houghton Mifflin Harcourt, 1930), 74.
27. "The tribal areas of Pakistan . . . present the perfect testing ground for a remote-controlled military strategy; it is a land set apart from its own country and mostly inaccessible to the international media and human rights groups, a place where violations of international law and civilian casualties go mostly uninvestigated. It is, in short, a black hole." Pir Zubair Shah, "My Drone War," *Foreign Policy*, February 27, 2012, cited in Derek Gregory, "Drones and Death in the Borderlands," in *Life in the Age of Drone Warfare* (Durham, NC: Duke University Press, 2017).
28. Kurt Halfyard, "Good Kill, Director Andrew Niccol Discusses The View From Above," *Screen Anarchy*, May 14, 2015, http://screenanarchy.com/2015/05/interview-good-kill-director-andrew-niccol-discusses-the-view-from-above.html.
29. In an interview with Justin Taylor, Brant describes Las Vegas as the simulacral counterpart of the operations of the Conex box: "A fake Paris, fake New York, and fake pyramid right next to a program for 'fake' combat—I couldn't resist." Justin Taylor, "Drone Plays: There's Something in the Air," *American Theater*, April 17, 2015, www.americantheatre.org/2015/04/27/drone-plays-theres-something-in-the-air/.
30. Megan Becker-Leckrone, *Julia Kristeva and Literary Theory* (New York: Palgrave, 2005), 63.
31. Clemens Apprich, "Remaking Media Practices: From Tactical Media to Post-Media," in: *Provocative Alloys: A Post-Media Anthology*, ed. Clemens Apprich, Josephine Berry Slater, Anthony Iles, and Oliver Lerone Schultz (Luneburg: PML Books/Mute, 2013), 122–140.
32. Adam Rothstein, *Drone* (London: Bloomsbury Press, 2015), xii.
33. The phrase "center of indetermination" is used to describe the subject in the work of Henri Bergson and subsequently reappears in the work of Gilles Deleuze. See chapter 1 of Henri Bergson, *Matter and Memory*, trans. N.M. Paul and W.S. Palmer (New York: Zone Books, 1988); and chapter 4 of Gilles Deleuze, *Cinema 1: The Movement-Image*, trans. Hugh Tomlinson and Barbara Habberjam (Minneapolis, MN: University of Minnesota Press, 1986).

CONCLUSION

1. US Department of Defense (DoD), "Network-Centric Warfare" (report to US Congress, July 27, 2001).
2. The intent was to urge Congress to take advantage of this extraordinary moment to reap unheard of rewards: "Just as the commercial sector required a

critical mass of connectivity, computers, and customers to successfully innovate with e-business solutions, DoD requires a similar critical mass of integrated communications and computing capability. The competitors who were first able to correctly identify the opportunity space provided by the Internet and e-business have been able to reap disproportionate rewards. The DoD seeks similar disproportionate advantages in future conflicts as we develop and implement a strategy for a network-centric transformation." DoD, "Network-Centric Warfare."

3. DoD, "Network-Centric Warfare," 3-1.

4. S. L. A. Marshall, *The Soldier's Load and the Mobility of a* Nation (Quantico, VA: Marine Corps Association, 1980), 75.

5. DoD, "Network-Centric Warfare."

6. Alex Williams and Nick Srnicek, "#ACCELERATE MANIFESTO for an Accelerationist Politics," part 7, May 14, 2013, http://criticallegalthinking.com/2013/05/14/accelerate-manifesto-for-an-accelerationist-politics/.

7. Steven Shaviro, *Post-Cinematic Affect* (London: Zero Books, 2010), 48.

8. Jodi Dean, *Democracy and Other Neoliberal Fantasies: Communicative Capitalism and Left Politics* (Durham, NC: Duke University Press, 2009).

9. Kate Conger and Dell Cameron, "Google Is Helping the Pentagon Build AI for Drones," Gizmodo, March 6, 2018, https://gizmodo.com/google-is-helping-the-pentagon-build-ai-for-drones-1823464533.

10. Manuel Castells, *Networks of Outrage and Hope: Social Movements in the Internet Age* (Cambridge, UK: Polity, 2015), 7–9.

11. James Der Derian, "The Desert of the Real and the Simulacrum of War," *International Affairs* 84, no. 5 (2008): 931–948.

12. "Gartner Says Almost 3 Million Personal and Commercial Drones Will Be Shipped in 2017," Gartner Newsroom, February 9, 2017, www.gartner.com/newsroom/id/3602317.

13. Slavoj Žižek, *The Plague of Fantasies* (London: Verso, 1997), 167.

14. Ulises Ali Mejias, *Off the Network: Disrupting the Digital World* (Minneapolis: University of Minnesota Press, 2013), 90.

15. Mejias, *Off the Network.*

16. Alexander R. Galloway and Eugene Thacker, *The Exploit: A Theory of Networks* (Minneapolis: University of Minnesota Press, 2007).

17. Digby Warde-Aldam, "Post-Surveillance: Suzanne Treister's Riposte to 'Post-Internet' Art," *Apollo: The International Art Magazine*, September 9, 2014, www.apollo-magazine.com/post-surveillance-suzanne-treisters-riposte-post-internet-art/.

18. Robin MacKay and Armen Avanessian, Introduction to *Accelerationist Reader* (Falmouth, UK: Urbanomic, 2014), 23.

19. Shane Denson, "Crazy Cameras, Discorrelated Images, and the Post-Perceptual Mediation of Post-Cinematic Affect," in *Post Cinema: Theorizing 21st Century Film*, ed. Shane Denson and Julia Leyda (New York: Reframe Books, 2016), https://reframe.sussex.ac.uk/post-cinema/2-5-denson/.

20. Lisa Parks, "Vertical Mediation and the U.S Drone War in the Horn of Africa," in *Life in the Age of Drone Warfare*, ed. Caren Kaplan (Durham, NC: Duke University Press, 2017), 140–141.

Bibliography

Abad-Santos, Alexander. "This 13-Year-Old Is Scared When the Sky Is Blue Because of Our Drones." *Atlantic*, October 29, 2013. Accessed March 5, 2019. www.theatlantic.com/politics/archive/2013/10/saddest-words-congresss-briefing-drone-strikes/354548.

Abelson, Hal. "Seductive Myths about Privacy." July 27, 2010. Accessed May 5, 2019. www.w3.org/2010/policy-ws/slides/09-Abelson-MIT.pdf.

Adey, Peter. *Aerial Life: Spaces, Mobility, Affect*. New York: Wiley, 2010.

Adorno, Theodor. *Minima Moralia*. [1951]. Accessed October 1, 2019. www.marxists.org/reference/archive/adorno/1951/mm/ch01.htm.

———. *Minima Moralia: Reflections on a Damaged Life*. New York: Verso, 2006.

Almond, Ian. "Derrida and the Secret of the Non-Secret: On Re-Spirtualising the Profane." *Literature & Theology* 17, no. 4 (December 2003): 457–471.

Anderson, Ben. "Facing the Future Enemy: US Counterinsurgency Doctrine and the Pre-insurgent." *Theory Culture & Society* 28, nos. 7–8 (2011): 216–240.

Animals and War: Confronting the Military-Animal Industrial Complex. Edited by Anthony J. Nocella II, Colin Salter, and Judy K. C. Bentley. Lanham, MD: Lexington Books, 2013.

"The Anthropoid Condition: Brían Hanrahan Interviews John Durham Peters." *Los Angeles Review of Books*, July 10, 2015. Accessed March 5, 2019. https://lareviewofbooks.org/article/the-anthropoid-condition-an-interview-with-john-durham-peters/.

Apprich, Clemens. "Remaking Media Practices: From Tactical Media to Post-Media." In *Provocative Alloys: A Post-Media Anthology*, edited by Clemens Apprich, Josephine Berry Slater, Anthony Iles, and Oliver Lerone Schultz, 122–141. Luneburg: PML Books/Mute, 2013.

Arnheim, Rudolf. *Film as Art*. London: Faber and Faber, 1933.

Assor, Nadav. "*Chariot* and *Ascender/Descender*." Digital videos and project description. 2014. Accessed March 5, 2019. www.nadassor.net/2014/10/chariot-ascender-descender-2014/.

Austin, J. L. *How to Do Things with Words*. Oxford: Oxford University Press, 1975.

Balázs, Béla. *Theory of Film: Character and Growth of a New Art*. Translated by Edith Bone. London: Dobson, 1952.

Balibar, Étienne. "Racism and Nationalism." In *Race, Nation, Class: Ambiguous Identities*, translated by Chris Turner, 37–68. New York: Verso, 1988.

Banner, Fiona. *Fiona Banner: Font*. Digital image. Frith Street Gallery. Accessed March 25, 2017. www.frithstreetgallery.com/exhibitions/fiona-banner-font.

Barthes, Roland. "Leaving the Movie Theater." In *The Rustle of Language*, 345–349. New York: Hill and Wang, 1986.

Baudry, Jean-Louis. "Ideological Effects of the Cinematographic Apparatus." In *Film Theory and Criticism*, edited by Leo Braudy and Marshall Cohen, 293–294. Oxford: Oxford University Press, 2004.

Becker, Jo, and Scott Shane. "Secret 'Kill List' Proves a Test of Obama's Principles and Will." *New York Times*, May 29, 2012. Accessed March 4, 2019. www.nytimes.com/2012/05/29/world/obamas-leadership-in-war-on-al-qaeda.html.

Becker-Leckrone, Megan. *Julia Kristeva and Literary Theory*. New York: Palgrave, 2005.

Beckman, Karen. "Telescopes, Transparency, and Torture: Trevor Paglen and the Politics of Exposure." *Art Journal* 66, no. 3 (2014): 62–67.

Benjamin, Medea, and Barbara Ehrenreich. *Drone Warfare: Killing by Remote Control*. London: Verso, 2013.

Berardi, Franco. "And Phenomenology of the End: Cognition and Sensibility in the Transition from Conjunctive to Connective Mode of Social Communication." PhD diss., Aalto University, School of Arts, Design and Architecture, Finland, 2014.Bergson, Henri. *Matter and Memory*. Translated by Nancy Margaret Paul and William Scott Palmer. New York: Zone Books, 1988.

Berlant, Lauren. *Cruel Optimism*. Chicago: University of Chicago Press, 2011.

Bignall, Simone. "Dismantling the Face: Pluralism and the Politics of Recognition." *Deleuze Studies* 6, no. 3 (2002): 389–410.

Bittel, Jason. "Want a Drone-Hunting License in Colorado? Get in Line." *Slate*, September 6, 2013. Accessed March 4, 2019. www.slate.com/blogs/future_tense/2013/09/06/deer_trail_co_sees_applications_for_nonexistent_drone_hunting_license.html.

Blanchot, Maurice. *The Instant of My Death*. Stanford, CA: Stanford University Press, 1998.

Blanchot, Maurice. *The Writing of Disaster*. Omaha: University of Nebraska Press, 2015.

Bonnington, Christina, and Spencer Ackerman. "Apple Rejects App That Tracks US Drone Strikes." *Wired*, August 30, 2012. Accessed June 15, 2019. www.wired.com/dangerroom/2012/08/drone-app/.

Borges, Jorge Luis. "On Exactitude in Science." In *Collected Fictions*, translated by Andrew Hurley, 325. New York: Penguin Books, 1999.

Brant, George. *Grounded*. London: Oberon Books, 2013.

Bräunert, Svea, and Meredith Malone. *To See Without Being Seen: Contemporary Art and Drone Warfare*. Chicago: University of Chicago Press, 2016.

Bridle, James. "The Art of Instagram: Dronestagram." *Blogs & Channel* (blog), May 8, 2015. Accessed March 5, 2019. www.tate.org.uk/context-comment/articles/art-of-instagram-dronestagram (link no longer active).

Bruhn, Jørgen. "What Is Mediality, and (How) Does It Matter? Theoretical Terms and Methodology." In *The Intermediality of Narrative Literature: Medialities Matter*, edited by Jørgen Bruhn, 13–40 (New York: Palgrave, 2016).

Bryant, Levi R. *The Democracy of Objects*. Ann Arbor: University of Michigan and Open Humanities Press, 2011.

Butler, Judith. "Ruled Out: Vocabularies of the Censor." In *Censorship and Silencing: Practices of Cultural Regulation*, edited by Robert C. Post, 247–258. Los Angeles: Getty Research Institute, 1998.

Calderone, Michael. "Robert Gibbs Told Not to Acknowledge Drone Program Exists as White House Press Secretary." *Huffington Post*, February 24, 2013. Accessed March 5, 2019. www.huffingtonpost.com/2013/02/24/robert-gibbs-drones-white-house_n_2753183.html.

Carrier, David. *Poussin's Paintings: A Study in Art-historical Methodology*. State College: Penn State Press, 1992.

Castells, Manuel. *Networks of Outrage and Hope: Social Movements in the Internet Age*. Cambridge, UK: Polity Press, 2015.

Cavallero, James, Stephan Sonnenberg, and Sarah Knuckey. *Living Under Drones: Death, Injury and Trauma to Civilians From US Drone Practices in Pakistan*. New York: NYU School of Law, Global Justice Clinic, September 25, 2012. Accessed March 5, 2019. https://law.stanford.edu/publications/living-under-drones-death-injury-and-trauma-to-civilians-from-us-drone-practices-in-pakistan.

Cebrowski, Arthur K., and John J. Garstka. "Network-Centric Warfare: Its Origin and Future." *Proceedings* 124, no. 1 (1998). Accessed March 4, 2019. www.usni.org/magazines/proceedings/1998/january/network-centric-warfare-its-origin-and-future.

Chamayou, Grégoire. "Psychopathologies of the Drone." In *A Theory of the Drone*, trans. Janet Lloyd, 106–113 New York: New Press, 2015.

Chamayou, Grégoire. *A Theory of the Drone*. Translated by Janet Lloyd. New York: New Press, 2015.

Chion, Michel. *Audio-Vision: Sound on Screen*. Translated by Claudia Gorbman. New York: Columbia University Press, 1994.

Clarke, David L. "On Being 'the Last Kantian in Nazi Germany': Dwelling with Animals after Levinas." In *Animal Acts: Configuring the Human in Western History*, edited by Jennifer Ham and Matthew Senior, 41–74. New York: Routledge, 1997.

Cloud, David S. "Anatomy of an Afghan War Tragedy." *LA Times*, April 10, 2011. Accessed March 4, 2019. http://articles.latimes.com/2011/apr/10/world/la-fg-afghanistan-drone-20110410.

Cohen, Kris R. "What Does the Photoblog Want?" *Media, Culture and Society* 27, no. 6 (2005): 883–901.

Coker, Christopher. *Future War*. Cambridge, UK: Polity Press, 2015.

Conger, Kate, and Dell Cameron. "Google Is Helping the Pentagon Build AI for Drones." Gizmodo, March 6, 2018. Accessed August 26, 2019. https://gizmodo.com/google-is-helping-the-pentagon-build-ai-for-drones-1823464533.

Cooper, Jill. *Animals in War*. London: Corgi, 1983.

Cotter, Holland. "Laura Poitras: Astro Noise' Examines Surveillance and the New Normal." *New York Times*, February 4, 2016.

Crandall, Jordan. "Ontologies of the Wayward Drone: A Salvage Operation." *CTheory*, 2011. Accessed March 5, 2019. http://ctheory.net/ctheory_wp/ontologies-of-the-wayward-drone-a-salvage-operation/.

———. "Un-manned." In *Manifestation & Materialization: UFO Drag Queen: Delirium*, 75–116. Atlanta, GA: Fort!/Da! Books, 2011.

Crogan, Patrick. "The Decomposition of Autonomy: Or, Drones and Global Technicity." Paper presented at Remote Control: Violence, Containment, Technology Conference, Utrecht, Netherlands, December 6–7, 2014. Accessed March 5, 2019. http://eprints.uwe.ac.uk/30847.

Cubitt, Sean. *Finite Media: Environmental Implications of Digital Technologies*. Durham, NC: Duke University Press, 2016.

Curcio, Seth. "Seeing Is Believing: An Interview with Trevor Paglen." *Daily Serving*, February 24, 2011. Accessed March 11, 2019. www.dailyserving.com/2011/02/interview-with-trevor-paglen/.

Currier, Cora. "How the Gov't Talks about a Drone Program It Won't Acknowledge Exists." *The Drone War* (blog), *ProPublica*, September 13, 2012. Accessed March 5, 2019. www.propublica.org/article/how-the-govt-talks-about-a-drone-program-it-wont-acknowledge.

Currier, Cora, and Justin Elliot. "The Drone War Doctrine We Still Know Nothing About." *ProPublica*, February 26, 2013. Accessed March 4, 2019. www.propublica.org/article/drone-war-doctrine-we-know-nothing-about.

Da Costa, Beatriz. "Pigeonblog 2006–08." *Beatriz Da Costa Blog & Project Hub* (blog), 2006. Accessed March 5, 2019. http://nideffer.net/shaniweb/pigeonblog.php.

Da Costa, Beatriz, with Cina Hazegh and Kevin Ponto. "Interspecies Coproduction in the Pursuit of Resistant Action." Project statement. Accessed February 12, 2015. http://nideffer.net/shaniweb/files/pigeonstatement.pdf.

De Villiers, Nicholas. "Leaving the Cinema: Metacinematic Cruising in Tsai Ming-liang's Goodbye, Dragon Inn." *Jump Cut* 50 (Spring 2008). Accessed March 5, 2019. www.ejumpcut.org/archive/jc50.2008/DragonInn/.

Dean, Jodi. *Democracy and Other Neoliberal Fantasies: Communicative Capitalism and Left Politics*. Durham, NC: Duke University Press, 2009.

DeLanda, Manuel. *Intensive Science & Virtual Philosophy*. London: Bloomsbury Press, 2013.

Delappe, Joseph. *Cowardly Drone*. Digital image. Accessed March 5, 2019. www.delappe.net/imaging/mq1-predator-drone---cowardly.

Deleuze, Gilles. *Cinema 1: The Movement Image*. Translated by Hugh Tomlinson and Barbara Habberjam. Minneapolis: University of Minnesota Press, 1986.

———. *The Fold: Leibniz and the Baroque*. London: Athlone Press, 1993.

———. "Immanence: A Life." In *Pure Immanence*, 25–34. New York: Zone Books, 2001.

———. "Postscript on Societies for Control." *October* 59 (Winter 1992): 3–7.

Deleuze, Gilles, and Félix Guattari. *A Thousand Plateaus: Capitalism and Schizophrenia*. Translated by Brian Massumi. Mineappolis: University of Minnesota Press, 1987.

Delmont, Matt. "Drone Encounters: Noor Behram, Omer Fast, and Visual Critiques of Drone Warfare." *American Quarterly* 65, no. 1 (March 2013): 193–202.

Denson, Shane. "Crazy Cameras, Discorrelated Images, and the Post-Perceptual Mediation of Post-Cinematic Affect." In *Post Cinema: Theorizing 21st Century Film*, edited by Shane Denson and Julia Leyda. New York: Reframe Books, 2016. Accessed March 1, 2019. http://reframe.sussex.ac.uk/post-cinema/2-5-denson/.

Derian, James Der. "The Desert of the Real and the Simulacrum of War." *International Affairs* 84, no. 5 (2008): 931–948.

Derrida, Jacques. "How to Avoid Speaking: Denials." In *Derrida and Negative Theology*, edited by Harold Coward and Toby Foshay, 73–136. Albany: State University of New York Press, 1992.

———. *The Truth in Painting*. Translated by Geoffrey Bennington. Chicago: University of Chicago Press, 1987.

Deutsche, Rosalyn. *Hiroshima after Iraq: Three Studies in Art and War*. New York: Columbia University Press, 2010.

Dillon, Michael, and Julian Reid. *The Liberal Way of War: Killing to Make Life Live*. New York: Routledge, 2009.

"Dolphins Inspire New Radar System." UCL Department of Electronic and Electrical Engineering. Accessed March 14, 2015. www.ucl.ac.uk/news/2013/oct/dolphins-inspire-new-radar-system.

Dorrian, Mark. "Drone Semiosis: Weaponry and Witnessing." *Cabinet*, Summer 2014, 48–55.

Drone May Have Spotted Bin Laden in 2000. Directed by David Ensor. March 17, 2004. Accessed March 4, 2019. www.cnn.com/2004/WORLD/asiapcf/03/17/predator.video.

"Drones Impact on Mental Health in Pakistan." *The Borgen Project* (blog), December 31, 2013, http://borgenproject.org/drones-impact-mental-health-pakistan/.

Drucker, Johanna. "Humanities Approach to Interface." *Culture Machine* 12, no. 9 (2011): 1–20.

Dundes, Alan. "Interpreting 'Little Red Riding Hood' Psychoanalytically." In *Little Red Riding Hood: A Casebook*, 16–51. Madison: University of Wisconsin Press, 1989.

Durham, John Peters. *The Marvelous Clouds: Towards a Philosophy of Elemental Media*. Chicago: University of Chicago Press, 2016.

Eckvahl, Colleen. "Drones Impact on Mental Health in Pakistan." *The Borgen Project* (blog), December 31, 2013. Accessed March 4, 2019. https://borgenproject.org/drones-impact-mental-health-pakistan.

Eder, Jens, and Charlotte Klonk. *Image Operations: Visual Media and Political Conflict*. Manchester, UK: Manchester University Press, 2016.

Eliot, T. S. "The Hollow Men." In *Selected Poems*, 75–80. London: Houghton Mifflin Harcourt, 1930.

Ensor, David. "Drone May Have Spotted bin Laden in 2000." CNN, March 17, 2004. Accessed February 12, 2019. www.cnn.com/2004/WORLD/asiapcf/03/17/predator.video/.

Erikson, Ruth. "Animal-Borne: Embodied Point of View and the Ethics of Identification." In *Cine-Ethics: Ethical Dimensions of Film Theory, Practice and Spectatorship*, edited by Jinhee Choi and Mattias Frey, 214–231. New York: Routledge, 2014.

Fair, C. Christian. "Ethical and Methodological Issues in Assessing Drones' Civilian Impacts in Pakistan." *Washington Post*, October 6, 2014. Accessed March 5, 2019. www.washingtonpost.com/blogs/monkey-cage/wp/2014/10/06/ethical-and-methodological-issues-in-assessing-drones-civilian-impacts-in-pakistan.

Farocki, Harun. "Phantom Images." *Public* 29 (2004). Accessed September 30, 2019. https://public.journals.yorku.ca/index.php/public/issue/view/1752/showToc.

Fisher, Mark. *Capitalist Realism: Is There No Alternative?* London: Zero Books, 2009.

Freud, Sigmund. "Fetishism." In *The Complete Works of Sigmund Freud*, 21:152–158. London: Hogarth, 1927.

———. "Inhibitions, Symptoms and Anxiety." In *The Complete Works of Sigmund Freud*, 20:75–176. London: Hogarth, 1926.

———. *The Standard Edition of the Complete Psychological Works of Sigmund Freud*. Edited by James Strachey. Vol. 21, *1927–1931*. New York: W. W. Norton, 1961.

Friedersdorf, Conor. "Every Person Is Afraid of the Drones: The Strikes' Effect on Life in Pakistan." *Atlantic*, September 25, 2012. Accessed March 4, 2019. www.theatlantic.com/international/archive/2012/09/every-person-is-afraid-of-the-drones-the-strikes-effect-on-life-in-pakistan/262814/.

Galloway, Alexander R., and Eugene Thacker. *The Exploit: A Theory of Networks*. Minneapolis: University of Minnesota Press, 2007.

"Gartner Says Almost 3 Million Personal and Commercial Drones Will Be Shipped in 2017." Gartner Newsroom, February 9, 2017. Accessed March 5, 2019. www.gartner.com/newsroom/id/3602317.

Gibson, William. *Spook Country*. New York: Penguin, 2007.

Gitelman, Lisa, and Virginia Jackson. Introduction to *"Raw Data" Is an Oxymoron*. Cambridge, MA: MIT Press, 2013.

Goodman, Nelson. *Ways of World-Making*. Indianapolis, IN: Hackett Publishing, 1978.

Goulimari, Pelagia. *Literary Criticism and Theory: From Plato to Postcolonialism*. London: Routledge, 2014.

Greenwald, Glenn. "ACLU Sues Obama Administration over Assassination Secrecy." *Salon*, February 2, 2012. Accessed March 5, 2019. www.salon.com/2012/02/02/aclu_sues_obama_administration_over_assassination_secrecy/.

———. "Obama DOJ Again Refuses to Tell a Court Whether CIA Drone Program Even Exists." *Guardian*, February 14, 2013. Accessed March 5, 2019. www.theguardian.com/commentisfree/2013/feb/14/cia-aclu-drone-secrecy.

Greenwald, Glenn, and Jeremy Scahill. "The NSA's Secret Role in the U.S. Assassination Program." *Intercept*, February 10, 2014. Accessed March 4, 2019. https://firstlook.org/theintercept/2014/02/10/the-nsas-secret-role.

Gregory, Derek. "Dirty Dancing: Drones and Death in the Borderlands." In *Life in the Age of Drone Warfare*, 25–58. Durham, NC: Duke University Press, 2017.

———. "The Everywhere War." *Geographic Journal* 177, no. 3 (September 2011): 238–250.

———. "From a View to a Kill: Drones and Late Modern Warfare." *Theory, Culture and Society* 28, no. 88 (2011): 188–215.

———. "Targeted Killings and Signature Strikes." *Geographical Imaginations: War, Spaces and Bodies*, November 6, 2012. Accessed March 4, 2019. http://geographicalimaginations.com/2012/11/06/targeted-killings-and-signature-strikes.

———. "Theory of the Drone 3: Killing Grounds." *Geographical Imaginations: War, Spaces and Bodies*, July 29, 2013. Accessed March 5, 2019. http://geographicalimaginations.com/2013/07/29/theory-of-the-drone-3-killing-grounds.

Groys, Boris. "Art, Technology, and Humanism." *E-Flux* 82 (May 2017). Accessed March 4, 2019. www.e-flux.com/journal/82/127763/art-technology-and-humanism/.

———. "The Politics of Installation." *E-Flux* 02 (January 2009). Accessed May 5, 2018. www.e-flux.com/journal/02/68504/politics-of-installation/

Halfyard, Kurt. "Interview: Good Kill, Director Andrew Niccol Discusses The View From Above." *Screen Anarchy*, May 14, 2015. Accessed March 5, 2019. http://screenanarchy.com/2015/05/interview-good-kill-director-andrew-niccol-discusses-the-view-from-above.html.

Hambling, David. "Cloak of Light Makes Drone Invisible?" *Wired*, May 9, 2008. Accessed March 5, 2019. www.wired.com/2008/05/invisible-drone.

———. "Game Controllers Driving Drones, Nukes." *Wired*, July 17, 2008. Accessed March 4, 2019. www.wired.com/2008/07/wargames/.

———. "Silence of the Drones: How to Quiet That Annoying Aerial Buzz." *New Scientist*, June 10, 2015. Accessed March 5, 2019. www.newscientist.com/article/dn27696-silence-of-the-drones-how-to-quiet-that-annoying-aerial-buzz.

Hand, Martin. *Ubiquitous Photography*. Cambridge, UK: Polity Press, 2012.

Haraway, Donna Jeanne. *When Species Meet*. Minneapolis: University of Minnesota Press, 2008.

Harrison, William Jerome. *A History of Photography*. London: Percy Lund and, 1897.

Hartman, Nathaniel. "Asm_frag." Accessed March 4, 2019. http://nathanielhartman.com/portfolio/asm_frag/.

Harwood, Graham. "Endless War: On the Database Structure of Armed Conflict." *Rhizome*, March 17, 2014. Accessed March 4, 2019. http://rhizome.org/editorial/2014/mar/17/endless-war-database-structure-armed-conflict/.

Hastings, Michael. "The Rise of the Killer Drones: How America Goes to War in Secret." *Rolling Stone*, April 16, 2012. Accessed March 5, 2019. www

.rollingstone.com/politics/news/the-rise-of-the-killer-drones-how-america-goes-to-war-in-secret-20120416.

Hayles, Katherine. *How We Think: Digital Media and Contemporary Technogenesis*. Chicago: University of Chicago Press, 2012.

Heddaya, Mostafa. "Landscape Acquired: A Dialogue with Scott Patrick Wiener." Hyperallergic, March 7, 2014. Accessed March 5, 2019. http://hyperallergic.com/113408/landscape-acquired-a-dialogue-with-scott-patrick-wiener.

Heidenreich, Stefan. "Freeportism as Style and Ideology: Post-Internet and Speculative Realism, Part I." *E-Flux* 71 (March 2016). Accessed March 5, 2019. www.e-flux.com/journal/71/60521/freeportism-as-style-and-ideology-post-internet-and-speculative-realism-part-i/.

Henn, Steve. "If There's Privacy in the Digital Age, It Has a New Definition." *All Tech Considered Blog: Policy*, March 3, 2014. Accessed March 5, 2019. www.npr.org/sections/alltechconsidered/2014/03/03/285334820/if-theres-privacy-in-the-digital-age-it-has-a-new-definition.

Hersch, Seymour M. "Rainmaking Is Used as Weapon by U.S." *New York Times*, July 3, 1972. Accessed March 5, 2019. www.nytimes.com/1972/07/03/archives/rainmaking-is-used-as-weapon-by-us-cloudseeding-in-indochina-is.html.

Hill, Mike. "Ecologies of War: Dispatch from the Aerial Empire." In *Telemorphosis: Theory in the Era of Climate Change*, Vol. 1, edited by Tom Cohen. Ann Arbor, MI: Open Humanities Press, 2012. Accessed March 4, 2019. https://quod.lib.umich.edu/o/ohp/10539563.0001.001/1:13/--telemorphosis-theory-in-the-era-of-climate-change-vol-1?rgn=div1;view=fulltext.

Holert, Tom. "Sensorship: The Seen Unseen of Drone Warfare." In *Image Operations: Visual Media and Political Context*, edited by Jen Elder and Charlotte Klonk, 101–117. Manchester, UK: Manchester University Press, 2017.

Holt, Nancy. "Sun Tunnels." In *Theories and Documents of Contemporary Art: A Sourcebook of Artists' Writings*, edited by Peter Selz and Kristen Stiles, 536–540. Chicago: University of Chicago Press, 2012.

Hong, Guo-Juin. "Theatrics of Cruising: Bath Houses and Movie Houses in Tsai Ming-liang's Films." In *Queer Sinophone Cultures*, edited by Howard Chiang and Ari Larissa Heinrich, 149–159. New York: Routledge, 2013.

Howard, Ian P., and Brian J. Rogers. *Binocular Vision and Stereopsis*. New York: Oxford University Press, 1995.

Hoyt, Kate Drazner. "Ethics of Network Subjectivity." *Technoculture* 4 (2014). Accessed 2014. https://tcjournal.org/vol4/hoyt.

Hsu, L. Y., T. Lepetit, and B. Kante. "Extremely Thin Dielectric Metasurface for Carpet Cloaking." *PIER: Progress in Electromagnetic Research* 152 (2015): 33–40.

Hull, Dana, and Ian King. "The Car Windshield Is Turning into a Computer Screen." Bloomberg, January 22, 2015. Accessed March 5, 2019. www.bloomberg.com/news/articles/2015-01-22/the-car-windshield-is-turning-into-a-computer-screen.

Hussain, Nasser. "The Sound of Terror: Phenomenology of a Drone Strike." *Boston Review*, October 16, 2013. Accessed March 5, 2019. http://bostonreview.net/world/hussain-drone-phenomenology.

"Intelligence Analysis," IBM. Accessed July 1, 2016, www.ibm.com/security/intelligence-analysis.

"IWM Talk: Covert War and Colliding Cultures by Mahwish Chishty," Imperial War Museum. Accessed March 5, 2019. https://imperialwarmuseum.wordpress.com/2016/12/20/iwm-talk-covert-war-and-colliding-cultures-by-mahwish-chishty/.

Jaeger, Suzanne M. "Embodiment and Presence: The Ontology of Presence Reconsidered." In *Staging Philosophy: Intersections of Theater, Performance and Philosophy*, edited by David Krasner and David Z. Saltz, 122–141. Ann Arbor: University of Michigan Press, 2006.

James, William. *Principles of Psychology*. New York: Dover Publications, [1890] 1950.

Kalnins, Karlis. "'Locative Media.'" Presented at Art and Communication Festival, Riga, Latvia.

Kaplan, Caren. "Drone-O-Rama." In *Life in the Age of Drone Warfare*, edited by Lisa Parks. Durham, NC: Duke University Press, 2017.

———. "Everything Is Connected: Aerial Perspectives, the Revolution in Military Affairs, and Digital Culture." *HASTAC*, August 18, 2009. Accessed March 4, 2019. www.hastac.org/electronic-techtonics/caren-kaplan-everything-connected-aerial-perspectives-revolution-military.

Kattenbelt, Chiel. "Intermediality in Theatre and Performance: Definitions, Perceptions and Medial Relationships." *Culture, Language and Representation: Cultural Studies Journal of Universitat Jaume I* 6 (2008): 19–29.

Klein, Melanie. "A Contribution to the Psychogenesis of Manic-Depressive States." *International Journal of Psycho-Analysis* 16 (1935): 145–174.

Kozol, Wendy. *Distant Wars Visible: The Ambivalence of Witnessing*. Minneapolis: University of Minnesota Press, 2014.

Krauss, Rosalind. *A Voyage on the North Sea: Art in the Age of the Post-Medium Condition*. London: Thames & Hudson, 1999.

Kravets, David. "CIA Refuses to Acknowledge Drone Targeted Killings." *Wired*, August 13, 2013. Accessed March 5, 2019. www.wired.com/2013/08/drone-targeted-killings/.

Krebs, Valdis. "Mapping Networks of Terrorist Cells." *Connections* 24, no. 3 (2002): 43–52.

Lane, Frank W. "Pigeons as Birds of War." *Flight: The Aircraft and Engineer*, October 21, 1943.

Latour, Bruno. "Spheres and Networks: Two Ways to Reinterpret Globalization." Lecture delivered at Harvard Graduate School of Design, Cambridge, MA, February 17, 2009. Accessed March 4, 2019. www.bruno-latour.fr/sites/default/files/115-SPACE-HARVARD-GB.pdf.

Lea, Stephen E. G., and Winard H. Dittrich. "What Do Birds See in Moving Video Images?" In *Picture Perception in Animals*, edited by Joel Fagot, 143–180. Philadelphia, PA: Taylor and Francis, 2000.

Leighton, T.G., G.H. Chua, P.R. White, K.F. Tong, H.D. Griffiths, and D.J. Daniels. "Radar Clutter Suppression and Target Discrimination Using Twin Inverted Pulses." *Proceedings of the Royal Academy* A469, no. 2160 (December 8, 2013). Accessed March 5, 2019. https://doi.org/10.1098/rspa.2013.0512.

Lenoir, Tim. Foreword to *New Philosophy for New Media*, by Marc Hansen. Cambridge, MA: MIT Press, 2004.

Leveringhaus, Alex, and Tjerk De Greef. "Keeping the Human 'In the Loop': A Qualified Defense of Autonomous Weapons." In *Precision Strike Warfare and International Intervention: Strategic, Ethico-Legal, Decisional Implications*, edited by Mike Aaronson, Wali Aslam, Tom Dyson, and Regina Rauxloh, 206–224. New York: Routledge, 2014.

Levinas, Emmanuel. *Ethics and Infinity: Conversations with Philipe Nemo.* Translated by Richard A. Cohen. Pittsburgh, PA: Duquesne University, 1995.

Lewis, Anthony. "Distant Emotions: 'Good Kill' and 'Slow West.'" *New Yorker*, May 8, 2015. Accessed March 5, 2019. www.newyorker.com/magazine/2015/05/18/distant-emotions.

Lippit, Akira. *Electric Animal: Towards a Rhetoric of Wildlife.* Minneapolis: University of Minnesota Press, 2008.

Liu, Alan. *The Laws of Cool: Knowledge Work and the Culture of Information.* Chicago: University of Chicago Press, 2004.

Mackay, Robin, and Armen Avanessian. *The Accelerationist Reader.* London: Urbanomic, 2014.

———. *#Accelerate: The Accelerationist Reader.* Falmouth, UK: Urbanomic Media, 2017.

Mackey, Robert. "Artists Try to Prick Conscience of Drone Operators with Giant Portrait of Orphan in Pakistani Field." *New York Times*, April 9, 2014. Accessed March 5, 2019. https://thelede.blogs.nytimes.com/2014/04/09/artists-try-to-prick-the-conscience-of-drone-operators-with-giant-portrait-of-orphan-in-pakistani-field.

Mandalaparthy, Nikhil. *UNEP Guide for Energy Efficiency and Renewable Energy Laws.* Nairobi, Kenya: United Nations Environmental Programme, 2015.

Marc, Franz. "How Does a Horse See the World." In *Theories of Modern Art: A Source Book by Artists and Critics*, edited by Herschel Browning Chipp and Peter Howard Selz, 178–179. Berkeley: University of California Press, 1968.

———. "Marc to Maria, 12 April 1915." In *Letters from the War: Franz Marc*, edited by Klaus Lankheit and Uwe Steffen, 46. Bern, Switzerland: Peter Lang, 1992.

Marinetti, F. T. "The Founding and Manifesto of Futurism." In *Futurism: An Anthology*, edited by Lawrence Rainey, Christine Poggi, and Laura Wittman, 49–53. New Haven, CT: Yale University Press, 2009.

Marshall, S. L. A. *The Soldier's Load and the Mobility of a Nation.* Quantico, VA: Marine Corps Association, 1980.

Martinussen, Einar Sneve, Jørn Knutsen, and Timo Arnall. "Satellite Lamps." *Kairos: A Journal of Rhetoric, Technology and Pedagogy* 19, no. 1 (Fall 2014). Accessed March 5, 2019. http://kairos.technorhetoric.net/19.1/inventio/martinussen-et-al/index.html.

Mbembe, Achille. "Necropolitics." *Public Culture* 15, no. 1 (Winter 2003): 11–40.

McGarry, Brendan. "Special Forces, Marines Embrace Palantir Software." *DefenseTech*, July 1, 2013. Accessed March 4, 2019. www.defensetech.org/2013/07/01/special-forces-marines-embrace-palantir-software.

McLuhan, Marshall. "The Gadget Lover: Narcissus as Narcosis." In *Understanding Media: The Extensions of Man*, edited by Terrence W. Gordon, 45–52. Corte Madero, CA: Gingko Press, 1994.

McNeal, Gregory S. "Are Targeted Killings Unlawful? A Case Study in Empirical Claims Without Empirical Evidence." In *Targeted Killings: Law and Morality in an Asymmetrical World*, edited by Claire Finkelstein, Jens David Ohlin, and Andrew Altman, 326–346. Oxford: Oxford University Press, 2012.

Mejias, Ulises Ali. *Off the Network: Disrupting the Digital World*. Minneapolis: University of Minnesota, 2013.

Miller, Greg. "Proposal to Give Federal Judges a Role in Drone Strikes Faces Hurdles." *Washington Post*, February 8, 2013. Accessed March 4, 2019. www.washingtonpost.com/world/national-security/proposal-to-give-federal-judges-a-role-in-drone-strikes-faces-hurdles/2013/02/08/66f53508-721a-11e2-8b8d-e0b59a1b8e2a_story.html.

Moholy-Nagy, László. "Directness of the Mind; Detours of Technology." Edited by Sibyl Moholy-Nagy. In *Moholy-Nagy*, edited by Richard Kostelanetz, 35–38. New York: Praeger, 1970.

Müller, Heiner. *Rotwelsch*. Berlin: Merve, 1984.

Nancy, Jean-Luc. *Dis-enclosure: The Deconstruction of Christianity*. Translated by, Bettina Bergo, Michael Bradley Smith, and Gabriel Malenfant. New York: Fordham University Press, 2009.

Nixon, Rob. *Slow Violence and the Environmentalism of the Poor*. Cambridge, MA: Harvard University Press, 2013.

#NotABugSplat. "About." April 6, 2014. Accessed March 6, 2019. https://notabugsplat.com/about.

"Obama War Chiefs Widen Drone Kill Box." *Washington Post*, December 3, 2013. Accessed August 20, 2019. www.washingtontimes.com/news/2013/dec/3/pentagon-gives-more-room-for-deadly-error-in-drone/.

Ottinger, Richard. "Section Two: Energy Resource Options." In *UNEP Guide for Energy Efficiency and Renewable Energy Laws*, 18–20. Nairobi, Kenya: United Nations Environment Programme (UN Environment), 2016.

Özkul, Didem. "Location as a Sense of Place: Everyday Life, Mobile and Spatial Practices in Urban Spaces." In *Mobility and Locative Media: Mobile Communication in Hybrid Spaces*, edited by Adriana De Souza E Silva and Mimi Sheller, 101–116. New York: Routledge, 2014.

Parikka, Jussi. *A Geology of Media*. Minneapolis: University of Minnesota Press, 2015.

Parks, Lisa. "Vertical Mediation and the U.S. Drone War in the Horn of Africa." In *Life in the Age of Drone Warfare*, edited by Caren Kaplan, 134–158. Durham, NC: Duke University Press, 2017.

Pavis, Patrice. "The State of Current Theatre Research." *Applied Semiotics/Sémiotique Appliquée* 1, no. 3 (1997): 203–230.

Pearl, Mike. "The #NotaBugSplat Art Piece in Pakistan Won't Be Making Drone Pilots Feel Empathy." *Vice News*, April 8, 2014. Accessed March 5,

2019. www.vice.com/en_uk/article/this-giant-art-piece-in-pakistan-wont-be-making-drone-pilots-feel-empathy.

Peters, John Durham. *The Marvelous Clouds: Toward a Philosophy of Elemental Media*. Chicago: University of Chicago Press, 2014.

Pettman, Dominic. *Infinite Distraction*. New York: Polity, 2015.

Pfaller, Robert. *Interpassivity: The Aesthetics of Delegated Enjoyment*. Edinburgh: Edinburgh University Press, 2017.

———. "Little Gestures of Disappearance: Interpassivity and the Theory of Ritual." *European Journal of Psychoanalysis* 16 (Spring 2003): 3–16.

Pleming, Clemency. "The Victorian Photographer Who Tried to Decipher Assyrian Tablets." *Oxford Arts Blog*, February 5, 2015, www.ox.ac.uk/news/arts-blog/victorian-photographer-who-tried-decipher-assyrian-tablets.

Porter, Gareth. "How McChrystal and Petraeus Built an Indiscriminate Killing Machine." Truthout, September 26, 2011. Accessed March 4, 2019. www.truth-out.org/news/item/3588:how-mcchrystal-and-petraeus-built-an-indiscriminate-killing-machine.

Poster, Mark. *The Mode of Information: Poststructuralism and Social Context*. Chicago: University of Chicago Press, 1990.

Power, Matthew. "Confessions of a Drone Warrior." *GQ*, October 23, 2013. Accessed March 4, 2019. www.gq.com/story/drone-uav-pilot-assassination.

Pugliese, Joseph. "Drone Casino Mimesis: Telewarfare and Civil Militarization." *Journal of Sociology* 52, no. 3 (2016): 500–521.

———. *State Violence and the Execution of Law Biopolitical Caesurae of Torture, Black Sites, Drones*. Abingdon, UK: Routledge, 2013.

Ranciére, Jacques. *The Politics of Aesthetics: The Distribution of the Sensible*. New York: Continuum, 2004.

Reprieve, "You Never Die Twice: Multiple Kills in the US Drone Program." November 24, 2014. Accessed March 4, 2019. https://reprieve.org/wp-content/uploads/2014_11_24_PUB-You-Never-Die-Twice-Multiple-Kills-in-the-US-Drone-Program-1.pdf.

Ressler, Steve. "Social Network Analysis as an Approach to Combat Terrorism: Past, Present, and Future Research." *Homeland Security Affairs* 2, no. 8 (July 2006). Accessed March 4, 2019. www.hsaj.org/articles/171.

Richardson, Ingrid. "Mobile Technosoma: Some Phenomenological Reflections on Itinerant Media Devices." *Fibre Culture* 6 (2005). Accessed March 5, 2019. http://six.fibreculturejournal.org/fcj-032-mobile-technosoma-some-phenomenological-reflections-on-itinerant-media-devices.

Rieder, Bernhard. "ORDER BY Column_name: The Relational Database as Pervasive Cultural Form." Paper presented at Lived Logics of Database Machinery Conference, London. 2012.

Rodowick, David Norman. *The Virtual Life of Film*. Cambridge, MA: Harvard University Press, 2007.

Rohde, David. "The Drone Wars." *Reuters Magazine*, January 26, 2012. Accessed March 5, 2019. www.reuters.com/article/us-david-rohde-drone-wars-idUSTRE80P11I20120126.

Romanes, George John. *Mental Evolution in Animals: With a Posthumous Essay on Instinct by Charles Darwin*. New York: D. Appleton and Company, 1884.

———. "Mental Evolution in Animals: With a Posthumous Essay on Instinct by Charles Darwin." In *Principles of Psychology*, by William James, 95–96. New York: Dover Publications, [1890] 1950.

Rothstein, Adam. *Drone*. London: Bloomsbury Press, 2015.

Rueb, Terri. "Restless: Locative Media as Generative Displacement." In *Mobility and Locative Media: Mobile Communication in Hybrid Spaces*, edited by Adriana De Souza E Silva and Mimi Sheller, 241–258. New York: Routledge, 2014.

Saifi, Sophia. "Not a 'Bug Splat': Artists Give Drone Victims a Face in Pakistan." CNN, April 9, 2014. Accessed November 20, 2018.. www.cnn.com/2014/04/09/world/asia/pakistan-drones-not-a-bug-splat.

Sandvig, Christian. "Seeing the Sort: The Aesthetic and Industrial Defense of 'The Algorithm.'" *Media-N: Journal of New Media Caucus*, November 22, 2014. Accessed March 4, 2019. http://median.newmediacaucus.org/art-infrastructures-information/seeing-the-sort-the-aesthetic-and-industrial-defense-of-the-algorithm.

Sankin, Aaron. "The Loneliness of the Long Distance Drone Pilot." *Kernel*, February 1, 2015. Accessed March 5, 2019. http://kernelmag.dailydot.com/issue-sections/headline-story/11553/air-force-drone-pilot-problem.

Schmid, Alex P. "The Definition of Terrorism." In *Routledge Handbook of Terrorism*, 39–99. New York: Routledge, 2011.

Schmitt, Carl. *The Concept of the Political*. Translated by George Schwab. Chicago: University of Chicago Press, 2007.

Shachtman, Noah. "Military 'Not Quite Sure' How Drone Cockpits Got Infected." *Wired*, October 19, 2011. Accessed March 5, 2019. www.wired.com/2011/10/military-not-quite-sure-how-drone-cockpits-got-infected.

Shah, Pir Zubair. "'My Drone War' Foreign Policy, Feb. 27, 2012." Cited in "Drones and Death in the Borderlands," in *Life in the Age of Drone Warfare*, by Derek Gregory, 25–58. Durham, NC: Duke University Press, 2017.

Shaviro, Steven. "Accelerationist Aesthetics: Necessary Inefficiency in Times of Real Subsumption." *E-Flux* 46 (June 2013). Accessed March 4, 2019. www.e-flux.com/journal/46/60070/accelerationist-aesthetics-necessary-inefficiency-in-times-of-real-subsumption.

———. *Post-Cinematic Affect*. London: Zero Books, 2010.

Shell, Hanna Rose. "The Crucial Moment of Deception." *Cabinet*, Spring 2009. Accessed March 4, 2019. www.cabinetmagazine.org/issues/33/shell.php.

Sheller, Mimi. "Mobile Art: Out of Your Pocket." In *Routledge Companion to Mobile Media*, edited by Gerard Goggin and Larissa Hjorth, 197–205. New York: Routledge, 2014.

Silverman, Kaja, *The Miracle of Analogy: The History of Photography Part I*. Stanford, CA: Stanford University Press, 2015.

Siskel, Gene. "The Touch That Transcends Violence and Death." *Chicago Tribune*, November 11, 1973.

Sloterdijk, Peter. *Terror from the Air*. Translated by Amy Patton and Steve Corcoran. Cambridge, MA: MIT Press, 2009.

"Smartbird: Bird Flight Deciphered." Festo. Accessed March 2, 2016. www.festo.com/net/SupportPortal/Files/46270/Brosch_SmartBird_en_8s_RZ_110311_lo.pdf.

Speri, Alice. "Giant Art Installation in Pakistan Tells US Drone Operators People Aren't 'Bug Splat.'" *Vice News*, April 7, 2014. Accessed March 5, 2019. https://news.vice.com/article/giant-art-installation-in-pakistan-tells-us-drone-operators-people-arent-bug-splat.

Stewart, Garrett. *Closed Circuits: Screening Narrative Surveillance*. Chicago: University of Chicago Press, 2014.

———. "Digital Fatigue: Imaging War in Recent American Film." *Film Quarterly* 62, no. 4 (Summer 2009): 45–55.

Stoichita, Victor. *The Self-Aware Image: An Insight into Early Modern Meta-Painting*. Cambridge, UK: Cambridge University Press, 2001.

Sutherland, Dane. "Vaporents: Inhuman Orientations." *Culture Machine* 16 (2015).

Tayag, Yasmin. "Drones Can Be Rendered Invisible with New Teflon Cloaking Technology: By Reflecting Their Surroundings, UAVs Can Disappear." *Inverse Innovation*, September 8, 2015. Accessed March 5, 2019. www.inverse.com/article/5975-drones-can-be-rendered-invisible-with-new-teflon-cloaking-technology.

Taylor, Justin. "Drone Plays: There's Something in the Air." *American Theater*, April 17, 2015. Accessed March 5, 2019. www.americantheatre.org/2015/04/27/drone-plays-theres-something-in-the-air.

"Threat Intelligence Analysis and Investigation." IBM Corporation, 2017. Accessed March 4, 2019. www.ibm.com/security/intelligence-analysis.

US Department of Defense (DoD). "Network-Centric Warfare." Report to Congress, July 27, 2001. www.dodccrp.org/files/ncw_report/report/ncw_main.pdf.

"US Drone Strikes Kill 28 Unknown People for Every Intended Target, New Reprieve Report Reveals." Reprieve, November 25, 2014. Accessed March 4, 2019. https://reprieve.org.uk/press/2014_11_25_us_drone_strikes_kill_28_each_target/.

Vierkant, Artie. "The Image Object Post-Internet." Accessed September 30, 2019. http://jstchillin.org/artie/pdf/The_Image_Object_Post-Internet_us.pdf.

Virilio, Paul. "Dromoscopy, or the Ecstasy of Enormities." *Wide Angle* 20, no. 3 (1998): 11–22.

———. *The Vision Machine*. Translated by Julie Rose. Bloomington: Indiana University Press, 1994.

Virno, Paolo. "PostFordist Semblance." *SubStance* 36, no. 1 (2007): 42–46.

Wangsness, T., and J. Franklin. "'Algorithm' and 'Formula.'" *Communications of the ACM* 9, no. 4 (1966): 243.

Warde-Aldam, Digby. "Post-Surveillance: Suzanne Treister's Riposte to 'Post-Internet' Art." *Apollo: The International Art Magazine*, September 2014. Accessed March 5, 2019. www.apollo-magazine.com/post-surveillance-suzanne-treisters-riposte-post-internet-art.

Weber, Jutta. "Keep Adding: On Kill Lists, Drone Warfare and the Politics of Databases." *Society and Space* 34, no. 1 (2016): 107–125.

Weber, Samuel. "Target of Opportunity: Networks, Netwar, and Narratives." *Grey Room* 15 (Spring 2014): 6–27.

———. *Targets of Opportunity: On the Militarization of Thinking*. New York: Fordham University Press, 2009.

Weil, Simone. *Gravity and Grace*. New York: Psychology Press, 2002.

Weinberger, Sharon. "How ESPN Taught the Pentagon to Handle a Deluge of Drone Data." *Popular Mechanics*, June 11, 2012. Accessed March 4, 2019. www.popularmechanics.com/technology/military/planes-uavs/how-the-pentagon-will-handle-its-deluge-of-drone-data-9600910.

Weingartner, James J. "Trophies of War: U.S. Troops and the Mutilation of Japanese War Dead, 1941–1945." *Pacific Historical Review* 61, no. 1 (February 1992): 53–67.

Whitaker, Raymond. "'Tungsten Bombs' Leave Israel's Victims with Mystery Wounds." *Independent*, January 18, 2009. Accessed March 5, 2019. www.independent.co.uk/news/world/middle-east/tungsten-bombs-leave-israels-victims-with-mystery-wounds-1418910.html.

Wilken, Rowan. "Locative Media: From Specialized Preoccupation to Mainstream Fascination." *Convergence: The International Journal of Research into New Media Technologies* 18, no. 3 (2012): 243–47. Accessed March 5, 2019. http://citeseerx.ist.psu.edu/viewdoc/download?doi=10.1.1.468.8295&rep=rep1&type=pdf.

Williams, Alex. "Escape Velocities." *E-Flux* 46 (June 2013). Accessed March 4, 2019. www.e-flux.com/journal/escape-velocities.

Williams, Alex, and Nick Srnicek. ""#ACCELERATE MANIFESTO for an Accelerationist Politics." Part 7, May 14, 2013. Accessed March 5, 2019. http://criticallegalthinking.com/2013/05/14/accelerate-manifesto-for-an-accelerationist-politics.

Woods, Chris. *Sudden Justice: Americas Secret Drone Wars*. Oxford: Oxford University Press, 2015.

Yusin, Jennifer. "Writing the Disaster: Testimony and The Instant of My Death." *Colloquy: Text, Theory, Critique* 10 (November 2015): 134–149.

Žižek, Slavoj. *The Plague of Fantasies*. London: Verso, 1997.

———. *The Sublime Object of Ideology*. London: Verso, 1989.

Illustrations

Index

Founded in 1893,
UNIVERSITY OF CALIFORNIA PRESS
publishes bold, progressive books and journals on topics in the arts, humanities, social sciences, and natural sciences—with a focus on social justice issues—that inspire thought and action among readers worldwide.

The UC PRESS FOUNDATION
raises funds to uphold the press's vital role as an independent, nonprofit publisher, and receives philanthropic support from a wide range of individuals and institutions—and from committed readers like you. To learn more, visit ucpress.edu/supportus.

Lightning Source UK Ltd.
Milton Keynes UK
UKHW012121110222
398568UK00001B/4